CHEMICAL ENGINEERING METHODS AND TECHNOLOGY SERIES

CHEMICALS AND RADIOACTIVE MATERIALS AND HUMAN DEVELOPMENT

CHEMICAL ENGINEERING METHODS AND TECHNOLOGY SERIES

Asphaltenes: Characterization, Properties and Applications
Jeremy A. Duncan
(Editor)
2010. ISBN: 978-1-60741-453-7

Asphaltenes: Characterization, Properties and Applications
Jeremy A. Duncan (Editor)
(Online Book)
ISBN: 978-1-61668-515-7

Bulk Materials: Research, Technology and Applications
Teodor Frías
and Ventura Maestas
(Editors)
2010. ISBN: 978-1-60692-963-6

Chromatography: Types, Techniques and Methods
Toma J. Quintin
(Editor)
2010. ISBN: 978-1-60876-316-0

Biomedical Chromatography
John T. Elwood
(Editor)
2010. ISBN: 978-1-60741-291-5

Biomedical Chromatography
John T. Elwood
(Editor)
2010. ISBN: 978-1-61668-470-9
(Online Book)

Flame Retardants: Functions, Properties and Safety
Paulo B. Merlani
2010. ISBN: 978-1-60741-501-5

Handbook of Membrane Research: Properties, Performance and Applications
Stephan V. Gorley
(Editor)
2010. ISBN: 978-1-60741-638-8

Non-Ionic Surfactants
Pierce L. Wendt
and Demario S. Hoysted
(Editors)
2010. ISBN: 978-1-60741-434-6

Bisphenol A and Phthalates: Uses, Health Effects and Environmental Risks
Bradley C. Vaughn
(Editor)
2010. ISBN: 978-1-60741-701-9

**Bisphenol A and Phthalates:
Uses, Health Effects and
Environmental Risks**
*Bradley C. Vaughn
(Editor)*
2010. ISBN: 978-1-61668-514-0
(Online Book)

**Chemicals and Radioactive
Materials and
Human Development**
*Rajiv K. Sinha, Rohit Sinha,
Shanu Sinha*
2010. ISBN: 978-1-61668-145-6

**Chemicals and Radioactive
Materials and
Human Development**
*Rajiv K. Sinha, Rohit Sinha,
Shanu Sinha*
2010. ISBN: 978-1-61668-455-6
(Online Book)

**Handbook of Hydrogels:
Properties, Preparation &
Applications**
*David B. Stein
(Editor)*
2010. ISBN: 978-1-60741-702-6

**Handbook of Hydrogels:
Properties,
Preparation & Applications**
David B. Stein
2010. ISBN: 978-1-61668-167-8
(Online Book)

**Fourier Transform Infrared
Spectroscopy: Developments,
Techniques and Applications**
*Oliver J. Rees
(Editor)*
2010. ISBN: 978-1-61668-835-6

**Handbook of Research
on Chemoinformatics
and Chemical Engineering**
A. K. Haghi (Editor)
2010. ISBN: 978-1-61668-504-1
(Online Book)

**Treatment of Tannery
Effluents by Membrane
Separation Technology**
*Sirshendu De,
Chandan Das,
Sunando DasGupta ,*
2010. ISBN: 978-1-60741-836-8

**Green Composites:
Properties, Design
and Life Cycle Assessment**
*François Willems
and Pieter Moens
(Editors)*
2010. ISBN: 978-1-60741-301-1

**Photocatalysis on Titania-
coated Electrode-less
Discharge Lamps**
*Vladimír Církva
and Hana Žabová*
2010. ISBN: 978-1-60876-842-4

Perovskites: Structure, Properties and Uses
Maxim Borowski (Editor)
2010. ISBN: 978-1-61668-525-6

Perovskites: Structure, Properties and Uses
Maxim Borowsk (Editor)
2010. ISBN: 978-1-61668-870-7
(Online Book)

Chemistry and Chemical Engineering Research Progress
A. K. Haghi (Editor)
2010. ISBN: 978-1-61668-503-4

Environmentally Harmonious Chemistry for the 21st Century
Masakazu Anpo and K. Mizuno (Editor)
2010. ISBN: 978-1-60876-428-0

Fluid Phase Behavior of Systems Involving High Molecular Weight Compounds and Supercritical Fluids
Pedro F. Arce and Martín Aznar
2010. ISBN: 978-1-61668-310-8

Preparation of Thin Film Pd Membranes for H2 Separation From Synthesis Gas and Detailed Design of a Permeability Testing Unit
M. Bientinesi, L. Petarca
2010. ISBN: 978-1-60876-538-6

CHEMICAL ENGINEERING METHODS AND TECHNOLOGY SERIES

CHEMICALS AND RADIOACTIVE MATERIALS AND HUMAN DEVELOPMENT

RAJIV K. SINHA
ROHIT SINHA
AND
SHANU SINHA

Nova Science Publishers, Inc.
New York

Copyright © 2010 by Nova Science Publishers, Inc.

All rights reserved. No part of this book may be reproduced, stored in a retrieval system or transmitted in any form or by any means: electronic, electrostatic, magnetic, tape, mechanical photocopying, recording or otherwise without the written permission of the Publisher.

For permission to use material from this book please contact us:
Telephone 631-231-7269; Fax 631-231-8175
Web Site: http://www.novapublishers.com

NOTICE TO THE READER

The Publisher has taken reasonable care in the preparation of this book, but makes no expressed or implied warranty of any kind and assumes no responsibility for any errors or omissions. No liability is assumed for incidental or consequential damages in connection with or arising out of information contained in this book. The Publisher shall not be liable for any special, consequential, or exemplary damages resulting, in whole or in part, from the readers' use of, or reliance upon, this material.

Independent verification should be sought for any data, advice or recommendations contained in this book. In addition, no responsibility is assumed by the publisher for any injury and/or damage to persons or property arising from any methods, products, instructions, ideas or otherwise contained in this publication.

This publication is designed to provide accurate and authoritative information with regard to the subject matter covered herein. It is sold with the clear understanding that the Publisher is not engaged in rendering legal or any other professional services. If legal or any other expert assistance is required, the services of a competent person should be sought. FROM A DECLARATION OF PARTICIPANTS JOINTLY ADOPTED BY A COMMITTEE OF THE AMERICAN BAR ASSOCIATION AND A COMMITTEE OF PUBLISHERS.

LIBRARY OF CONGRESS CATALOGING-IN-PUBLICATION DATA

Available Upon Request

ISBN: 978-1-61668-145-6

Published by Nova Science Publishers, Inc. ✚ New York

Contents

Preface		ix
Acknowdlegments		xiii
Chapter 1	Introduction	1
Chapter 2	Chemicals in the Service of Mankind: A Boon or Bane?	5
Chapter 3	Chemicals on the Poles: Omnipresent on Earth	7
Chapter 4	Chemically Contaminated Sites on Earth: Legacies of Maldevelopment	9
Chapter 5	Toxic Chemicals Used in Consumer Products of Everyday Use	11
Chapter 6	Chemicals Classified by their Potential Toxicity and Health Impacts on Human Beings	15
Chapter 7	Health Risk Assessment of Chemicals: The Demons of Development	19
Chapter 8	Chemicals in the Breathing Air: Their Potential Health Risks and Impacts	23
Chapter 9	Some Key Chemicals in the Air Pollutants and Health Impacts	29
Chapter 10	Chemicals in the Human Environment Due to Vehicular Pollution	33
Chapter 11	Chemicals in the Indoor Air: The Hidden Danger in Modern Homes	35
Chapter 12	Health Impacts of Chemicals in the Water	37
Chapter 13	Health Impact of Chemical Contamination of Food and Farm Soil Due to Widespread Use of Agrochemicals	41
Chapter 14	Synthetic Chemicals Used in Human Developmental Activities and their Potential Health Risks and Impacts	43

Chapter 15	Heavy Metals Used in Developmental Activities and Risks to Human Health	59
Chapter 16	Exposure to Chemicals: Mechanism of Entry into Human Body	67
Chapter 17	Fate of Chemicals in Human Body	71
Chapter 18	Epidemiological Studies of Chemicals Used in Developmental Activities	73
Chapter 19	Pharmaceutical Drugs and Chemicals: Mixed Blessing for Mankind	79
Chapter 20	Some Important Global Actions for Removal of Toxic Chemicals from Human Environment: The Green Chemistry Movement	87
Conclusions		107
Index		115

PREFACE

Hazardous chemicals and radioactive substances have permeated the human environment and the ecosystem. They are now in our basic life support systems- air, water, food and soil. Scientific investigations have shown over 350 'synthetic chemicals' in human environment. The matter of more serious concern is that human beings have become exposed to synthetic chemicals for which there has been no past evolutionary adaptation and experience. The chemicals are completely 'foreign' to all living organisms. Some synthetic chemicals mimic natural body hormones and send false messages. Others 'block the messages' by disruption and prevent true ones from getting through. Any chemical which interferes directly or indirectly with hormone function can scramble vital messages, derail development and undermine health. Industrial chemicals may disrupt normal function of body's 'hormonal system' causing reproductive and developmental abnormalities, neurological and immunological problems and cancer.

The World Health Organization (WHO) reports that 25 % of all preventable illnesses are directly caused by environmental pollution (chemical, radiological and biological) of the life-support systems. On global level, intoxication attributed to chemical pesticides have been estimated to be as high as 3 million cases of acute, severe poisoning annually, with as many or even more unreported cases mostly in the developing countries, and some 220,000 deaths. Any chemical, in the form of a liquid, dust, vapor, gas, aerosol or mist can enter the eye by dissolving in the liquid surrounding the eye. As eyes are richly supplied by blood vessels, many chemicals can penetrate the outer tissues and pass into the veins.

Modern life cannot be imagined without chemicals and some are crucial to our well-being. Production of goods of widespread societal use and consumption such as paper, plastic, leather, textile and even food incur heavy use of chemicals. From transportation to information technology, to the operation of electrical equipments and heavy machines and tools, and in entertainment, some chemicals are used. But some sustainable solution is imminent as it would affect our very survival on earth.

Science has helped us identifying the harmful chemicals specially the carcinogens, teratogens, immunotoxins and endocrine disruptors used in production process and their release into the human environment, making it possible to restrict or even eliminate their production and use. After the

Stockholm Conference on Human Environment in 1972 several nations banned the production of toxic synthetic chemicals through appropriate legislation. A Green Chemistry Movement (GCM) is also going across the world since 1990 to reduce or even eliminate the use of toxic chemicals in industrial production process and search for benign alternatives.

Scientists in the U.S. have identified a 'master gene' that controls the action of 50 other genes whose products protect the lungs against chemicals and pollutants in the environment. The master gene named as 'nrf2' is activated in response to environmental pollutants which then turns on numerous antioxidant and pollutant-detoxifying genes to protect the lungs from developing emphysema.

Keywords: *Exposure to Chemicals in Everyday Life; Chemical Residues in Food; Chemicals in Air, Water and Soil Adversely Affects Human Health; Household Hazardous Wastes-Poison in Homes; Human Disasters Due to Hazardous Chemicals; Radioactive Materials in Human Environment-Potential Risk to Civilization; People's Protest Against Hazardous Chemicals in Consumer Products; Biological Half-time of Chemicals in Human Body; Green Chemistry Movement to Eliminate Toxic Chemicals from Human Environment; Public Opposition Against Nuclear Power; Master Gene in Humans and its Response to Pollution*

ACKNOWLEDGMENTS

We are grateful to all those learned authors, editors and publishers of the books and journals whose papers and articles provided valuable informations on the subject and helped in the preparation of this volume. Their names have been duly referred in the list of references. And, to the best of our knowledge we have taken all care not to violate the copyrights of the learned authors but if that might have had happened unknowingly & untentionally, we all duly apologize to those learned authors and scientists. Our intention is to spread the knowledge about 'potential health impacts of environmental pollutants' among the global human society and to make aware the policy makers and developers to change the current pattern of development and find a more environmentally and socially sustainable alternatives. It has to be 'development without destruction' and development and environment must go hand in hand.

We especially acknowledge the publications of UNEP (United Nation Environment Program) and the WHO (World Health Organization) which provided immense informations on the subject.

Rajiv K. Sinha
Rohit Sinha &
Shanu Sinha

Chapter 1

INTRODUCTION

Despite significant improvements in pollution abatement and environmental remediation over the past several decades, billions of people around the world continue to live in unsafe and unhealthy physical environment with great risk to their health. The poor are disproportionately at greater risk. The World Health Organization (WHO) reports that 25 % of all preventable illnesses are directly caused by environmental factors, mostly due to the pollution (chemical, radiological and biological) of the life-support systems- air, water and soil. Estimates done by Harvard University (U.S.) research group (1996) suggest that premature death and illness as a result of major environmental health risks account for a fifth of the disease burden in the developing world. Of the 800,000 premature deaths attributed to urban air pollution every year about 65 % occur in Asia. (GEO, 2006). Local air pollution due to use of dirty cooking fuels causes perhaps 4 million premature deaths (mostly of young children under five) every year in developing countries. The Ohio State University Medical Center reported direct link between 'air pollution, obesity and type II diabetes'. Researches found that exposure to air pollution, over a period of 24 weeks, exaggerates 'insulin resistance' and adipose tissues inflammation. Another study implicated air pollution as a major adverse risk factor for 'cardiovascular effects', 'high blood pressure' and acute 'coronary syndromes' (Rajgopalan, 2009). The economic and health costs of local air pollution amount to over US $ 350 billion a year, or 6 % of the gross national product (GDP) of developing countries. (UNEP Report, 2003).

Chemical revolution began to unfold in the mid-1950s. Chemicals –the demons of development- are all around us, with tens of thousands currently in use and inevitably find their way into our bodies. They have become an intrinsic part of the lives of modern human society and are used in virtually all consumer products- cars, papers, plastics, textiles, electronics, building materials, food and medicine. Workers in mining, metallurgical, fertilizer, paint, plastic and automobile industries often face prolonged exposure to several hazardous chemicals. The persistent organic pollutants (POPs) like PCBs and DDT are found almost everywhere- in our food, soil, air, and water. Humans and wildlife around the world carry amounts in their body that is alarming. Recent studies have confirmed that in some countries of North, 40 to 65 % of women have levels of PCBs in their blood that are up to 5 times higher than recommended. Our babies receive their first dose while still in their mother's womb. More chemicals reach to them through their mother's milk. (UNEP Reports, 2002- 06).

Hazardous chemicals have permeated the human environment and the ecosystem. They are now in our basic life support systems- air, water, food and soil. Recent scientific investigations have shown that over 350 'synthetic chemicals' have been found in the humans (UNEP and WHO Report, 2002). What is the matter of more serious concern is that living organisms including the human beings, have become exposed to synthetic chemicals for which there has been no evolutionary adaptation and experience. The chemicals are completely 'foreign' to living organism. On global level, intoxication attributed to chemical pesticides have been estimated to be as high as 3 million cases of acute, severe poisoning annually, with as many or even more unreported cases mostly in the developing countries, and some 220,000 deaths. Industrial chemicals may disrupt normal function of body's 'hormonal system' causing reproductive and developmental abnormalities, neurological and immunological problems and cancer. The Mexican episode of babies born without brain (anenocephaly) is linked with the toxic and hazardous chemical waste dumps from the industries. (UNEP Report, 1992). Other medically active chemicals recently discovered are n-nitrosodimethylamine (NDMA) (a principal ingredient in rocket fuel), methyl tertiary butyl ether (a highly soluble gasoline additive), and phenolic compounds.

The urge to generate 'nuclear power' as a clean source of enormous energy from nuclear fuel uranium (which is of course proving to be 'curse in disguise' today) unleashed enormous amount of radioactive materials into the human environment. All radioactive substances emit 'harmful rays' which can pass through human body destroying the vital cells and genetic materials. They pose serious threat to human health, and the organs most sensitive to radiation are the 'reproductive organs' and the 'eyes'. All over the world the 'male sperm count' is declining and it is attributed to the background radiation caused by the millions of tones of radioactive wastes in the environment. Radon is a demonstrated cause of lung cancer. Underground uranium miners in the U.S. in 1950s who were exposed to radon suffered from lung cancer. The combination of radon and smoking is particularly deadly. The ingestion of radioactive products from the use of radioactive water in industries can have a somatic effect on human beings, causing malignant tumours, or chromosomal and gene (heredity materials) mutations that might affect the future generations. In the recent years, people all over the world have been up in arms against uranium mining and the nuclear power plants which generates radioactive wastes. The 'Greenpeace Society' and the 'Friend's Of Earth' have been particularly active in arousing public awareness against the ill effects of radioactive wastes on human health.

Scientists working on 'Environmental Health Sciences' in the U.S. have identified a 'master gene' that controls the action of 50 other genes whose products protect the lungs against environmental pollutants. The master gene named as 'nrf2' is activated in response to environmental pollutants which then turns on numerous antioxidant and pollutant-detoxifying genes to protect the lungs from developing emphysema. They indicated that the master gene nrf2 was also activated in response to an anti-cancer agent 'sulforaphane'.

Chapter 2

CHEMICALS IN THE SERVICE OF MANKIND: A BOON OR BANE ?

Modern human civilization is based on copious use of chemicals in every walk of life. There is no sector of human development activity which do not make use of some toxic chemicals and materials which eventually end up as a chemical waste. Life today, cannot be imagined without chemicals. We cannot get away from the chemicals, it is all around us - in the soil where we grow our food, in the food (as residual pesticides) we eat, in the air (as pollutant) we breathe, and in the water (as contaminant) we drink. Approximately 80,000 chemicals have been introduced into the human environment over the last 50 years and about 15,00 new ones are added each year. The global production of chemicals has increased from 1 million tones in 1930 to 400 million tones in 2000 (UNEP Report, 2001). Some chemicals are like 'necessary evil'. Chemical substances recorded in the European Union in 1980s, were over 100,000 of them (UNEP Report, 2004). Since 1950 there has been a tremendous increase in the production of 'organic chemicals' to satisfy our demands for consumer goods. Production of goods of widespread societal use and consumption such as paper, plastic, leather, textile and even food incur heavy use of chemicals and consequently discharge large amount of chemical waste and pollution. The DDT and the PCBs were produced at about 100 million pounds per year during the 1960s and 70s. However, they are banned now. Since 1950 there has also been a tremendous increase in the use of 'heavy metals' to satisfy our demands for consumer goods.

Many chemicals are crucial to our well-being. They are essential for human development and in everyday life. From transportation to information technology, to the operation of electrical equipments and heavy machines and tools, and in entertainment, some chemicals are used. They are there in life saving medicines, in the food preservatives, paints and plastics which has revolutionized the 21st century life-style. Unfortunately, their use eventually end up as hazardous waste. Ironically, chemicals that were developed to control diseases, increase food production, and improve our standard of living, have become a real threat to our very existence.

Chapter 3

CHEMICALS ON THE POLES: OMNIPRESENT ON EARTH

Hazardous chemicals and wastes do not remain confined to their place of origin. They are transported to long distances by the atmosphere. 16 pesticides used to control pests in the crop lands have been found in fogs which are inhaled by people normally. Some hazardous wastes and chemicals such as the 'persistent organic pollutants' (POPs) have even reached the poles in the Arctic and the Antarctica. Snow on the Swiss Alps holds DDT used for malaria control in the tropics. DDT and PCB's are present in the polar bears, seals and the penguin birds. The indigenous communities in the Arctic Circle carry PCBs used primarily as flame retardants in the South. Certain POPs in some Inuit societies of northern Canada, Alaska and Greenland is 10 to 20 times higher than in most temperate regions and they are at risk. Once in the Artic, they degrade very slowly and bio-accumulate. These POPs were originally released in the tropical and temperate environments some 40 – 50 years back.

POP's are more hazardous because they are highly bio-accumulative in the living systems. Although not soluble in water, they are readily absorbed in the fatty tissues, where they can become concentrated up to 70,000 times the background levels. They can cause cancer, allergies and hypersensitivity, damage the central and peripheral nervous system, cause reproductive disorders, birth defects, and disruption of the human immune system.(UNEP-DTIE, 2002).

Chapter 4

CHEMICALLY CONTAMINATED SITES ON EARTH: LEGACIES OF MALDEVELOPMENT

No corner of the world are free of the 'chemical sins and legacies' of modern development. There are tens of thousands of square kilometers of land on earth contaminated by dumping of hazardous chemical wastes. 40,000 contaminated sites in US, 55,000 in just six European countries and 7,800 in New Zealand have been reported (UNEP Report). Cleaning them up will require billions of dollars. There is serious risk of contamination of ground water in these regions. There were reports of anenocephaly (babies born without brain) from population living near hazardous waste dump sites in Mexico City which was closed decades ago. *The civilization is practically sitting at the mouth of the manmade 'volcano' dumped with chemical and nuclear wastes- the worst legacies of science.*

Chapter 5

TOXIC CHEMICALS USED IN CONSUMER PRODUCTS OF EVERYDAY USE

Many untested petrochemicals are used in a wide range of household products, including fabric softeners, air fresheners, perfumes and cosmetics, paper products and other consumer goods.

1) There are some chemicals in the hair dyes, hair removers, hair shampoos and shaving creams, toothpastes and mouthwash, room and body spray, soaps and detergents which we use everyday. The popular disinfectant traded in the name of 'Harpic' contains 'benzalkonium chloride' 0.6 % w/v and 'Ajax' contain 'bleach'. The anti-dandruff hair shampoo 'Head and Shoulder' contains 'benzyl alcohol', and 'methylchloroisothiazolinone', others contain 'pyrithione zinc' 1 % w/w. The hair conditioner 'Pantene' contains 'cyclopentasiloxane', 'steramidopropyl demethylamine', 'dimethicone benzyl alcohol', and 'penthenyl ethyl ether'.
2) The white paper which we use requires chemical bleaching by chlorine. Paper consumption worldwide produces several million tones of 'chemical cocktail' every year, so lethal that one drop in a swimming pool would kill a trout.
3) There are hazardous chemicals in the toys and furniture, in the inks, paints, polishes and adhesives, in solders and batteries, in the bleached papers, plastics, newspapers, nappies and sanitary towels.
4) There is a chemical in the 'correction fluid' which has potential to kill.

5) There is an explosive chemical in the safety air bags now used in all modern cars.
6) Carbon tetrachloride is used in dry-cleaning processes.
7) Heavy metal cadmium is present in plastics.
8) The highly toxic polychlorinated biphenyls (PCBs) are added to paints, copying and printing paper inks, adhesive and plastics to improve their flexibility.
9) All our paints and enamels used in homes and in automobiles contain dangerous chemicals like glycol, ether, ammonia, benzene and formaldehyde. They continue to give out toxic fumes at least for 7 years, significantly contributing to the indoor pollution in homes and institutions.
10) Water industries use several chemicals including toxic 'chlorine' in the water treatment process before municipal supply and also for the wastewater treatment before discharge into the environment. Evidences are accumulating that chlorination of water leads to formation of carcinogenic 'trihalomethanes' due to reaction with NOM (natural organic matters) in the water.
11) A wide variety of chemicals are now being used for the procurement of human food from their production in the farms to their preservation in the food grain storage facilities, and processing in the food industries. The agro-chemical industries produce over 100 different formulations of pesticides, insecticides, herbicides and fungicides used in food production and preservation. All canned and bottled foods contains chemicals as 'preservatives'. Heavy metal cadmium is present in food processing equipment, kitchenware enamels, pottery glazes and relatively high levels in the sea foods. Fish food contain generally higher levels of PCB's. High levels of PCB's were reported from the breakfast cereals in Sweden and Mexico as a result of contamination by 'packaging materials'.

Over 131 types of chemical pesticides used in agriculture and horticulture exists on Earth. They include chlorinated hydrocarbons, organophosphates, carbamates, synthetic organic pyrethroids, fumigants, herbicides, and rodenticides. Organochlorines e.g. DDT are of widespread use in the human society for non-agricultural purposes too.

Perfumes and Cosmetics : A Curse in Disguise

National Institute of Occupational Safety and Health in the US has reported 884 'neurotoxic' chemical compounds in the products of the perfumes, cosmetics and toiletries used by the women in her daily life. Researches done at Sweden found 'di-ethylhexyl phthalate' or other phthalates in 34 leading brands of cosmetics and toiletries including the Chanel No. 5, Christian Dior's Poison, Eternity from Calvin Klein and Tresor by Lancome, Tommy Girl Perfume, Impulse Body Spray, Nivea Deo Compact, Sure Ultra deodorant, and 4 hairsprays including Elnett Satin, Pantene Pro-V Extra Hold and Vidal Sassoon.

The phthalates which help prevent loss of fragarance has been found to cause 'genetical abnormalities' and damage to the reproductive systems including 'testicular cancer' and 'infertility' which have risen to tenfold in the past century. The phthalates could be absorbed into women's bloodstreams through the skin or inhalation. The European Commission is proposing to ban on the use in cosmetics of two of the most potent forms of phthalates which has been found to cause genetical abnormalities in up to 4 % of male babies. (UNEP Reports, 1992- 2006).

Chapter 6

CHEMICALS CLASSIFIED BY THEIR POTENTIAL TOXICITY AND HEALTH IMPACTS ON HUMAN BEINGS

Based on their toxicity effects following categories of chemicals are identified in environment-

ASPHYXIANT

A chemical that interferes with the oxygenation of the body tissues. It interferes with the body's ability to transport and use oxygen. Better-known examples are carbon monoxide and cyanides.

CARCINOGEN

A chemical that causes cancer. Several industrial chemicals are carcinogenic. Better known examples are pentachlorophenol (PCP), benzene, heavy metals asbestos (Ab), arsenic (As), lead (Pb), chromium (Cr).

MUTAGEN

A chemical that can cause permanent damage to the genetic material DNA in a cell. It leads to birth defects and heritable diseases. Better known examples are Bisphenol A, PCBs, arsenic etc.

TERATOGEN

A chemical that, if present in the bloodstream of a pregnant women, cross the placenta, affecting the developing fetus and resulting in structural or functional congenital abnormalities or cancer, in the child. These effects may not be observable until the child becomes an adult. Several industrial chemicals especially, nickel, methanol, ethylene dichloride dimethyl formamide and vinyl chloride are known teratogens. Thaliomide is another well known example. In the 1960s it caused many cases of 'phocomelia' in pregnant women. There was severe reduction in the limbs to the extent that hands and feet of the baby were attached directly to the body.

FETOTOXICANT

A chemical that adversely affects the developing fetus, resulting in low birth weight or stillbirth (the fetus dies before it is born). Better know example is again pentachlorophenol (PCP).

NEUROTOXINS

Chemicals that affect the brain and nerves are called neurotoxins. They lead to slowing down brain and nerve function leading to paralysis. Better examples are heavy metals mercury (Hg) especially methyl mercury, lead (Pb).

NEPHROTOXINS

Chemicals that affect the kidneys are called nephrotoxins. It can lead to sudden failure of the kidneys (acute renal failure), chronic renal failure, and also cancer of the kidney or bladder. Better known examples are heavy metals cadmium (Cd) and beryllium (Be).

HEPATOTOXINS

Chemicals that affect the liver are hepatotoxins. Most chemicals in human body are metabolized in liver, and many therefore have the potential to damage the liver cells. Possible short-term affects of chemicals on the liver are chemical hepatitis (inflammation of the liver cells), necrosis (cell death) and jaundice. Long-term effects lead to cirrhosis (scarring) of the liver and liver cancer. Better known examples are dimethylformamide,

IMMUNOTOXINS

Chemicals affecting the human immune system and impairing the immune response. Better known examples are the organochlorines- the polychlorinated biphenyls (PCBs) and the dioxins. Certain heavy metals are highly 'toxic' and provoke 'immune reactions' e.g. mercury (Hg), gold (Au), platinum (Pt), beryllium (Be), chromium (Cr), and nickel (Ni). Two main immune system impairment may be associated with exposure to the toxins- direct damage that cause 'immune suppression' which may render individual more susceptible to infections and cancer; and indirect response that cause 'immune sensitization' which develop allergic or hypersensitivity reactions to foreign antigens.

Chapter 7

HEALTH RISK ASSESSMENT OF CHEMICALS: THE DEMONS OF DEVELOPMENT

Toxic chemical poses serious risk to human health and environment at every stage – from production to use, and their transportation and treatment for safe disposal. They pose grave risks to human health at all times- during generation, storage as well as during disposal and it is a risk not only to the present generation but also the future generations as it affects the 'hereditary materials'. The products of industrial chemical revolution include many chemicals and chemical structures that can be considered 'foreign' to living organisms. Human beings (*Homo sapiens*) have become exposed to chemicals for which there has been no biological (immunological) adaptation to cope with. The organism may not get the chance to do anything if the compound is toxic and interferes with some steps in the organism's life-sustaining biochemical and physiological process going on in the body.

The environmental and human health consequences of the chemicals and wastes of our technological development were not understood or even recognized initially. It took years or decades for chronic effects to manifest themselves, and those cases were obscured by the fact that everyone is exposed to a wide number of chemicals.

The landmark episodes of human disasters caused by poisoning effects of methyl mercury, dioxins, DDT and PCB's alarmed the world about the potential dangers of the chemicals that were being piled up in the human ecosystem. According to United Nation Environment Program (UNEP) and the World Health Organization (WHO) nearly 3 million people suffer from 'acute pesticide poisoning' and some 10 to 20 thousands people die every year from it in the developing countries. US scientists predict that up to 20,000

Americans may die of cancer, each year, due to the low levels of 'residual pesticides' in the chemically grown food.

TOXIC EFFECTS OF CHEMICALS ON HUMANS AND THE FACTORS INFLUENCING TOXICITY

Toxicity of a chemical / substance may be defined as the substance's capacity to harm a living organism including human beings. A highly toxic substance will harm an organism even if very small amounts are present in the body. Conversely, a substance of low toxicity will not produce any effect unless the amount present in the body is very large. For a chemical to exert an effect, there must first be an exposure. Not even the most toxic chemicals will cause harm to an organism, including humans, if there is no exposure.

On a health hazard spectrum (HHS) of 0 – 3, a score of 3 represents a very high hazard to health, 2 represent a medium hazard and 1 is harmful to health. Factors that are taken into account to obtain this ranking include the extent of the material's toxic or poisonous nature and /or its lack of toxicity, and the evaluation of its tendency to cause, or not cause CANCER and /or BIRTH DEFECTS. It does not take into account exposure to the substance.

Factors that influence toxicity of a chemical are-

1. Quantity of the chemical absorbed (the dose) by the person exposed and the susceptibility of the exposed person;
2. Route of exposure e.g. inhalation, ingestion, or absorption through skin;
3. Duration and degree of exposure to the chemical and how often that exposure occurs; and
4. The type and severity of the injury caused whether permanent (irreversible) or reversible.

Gender, age and race are also factors influencing toxicity. Women with a greater relative proportion of body fat (potential site of storage of chemicals in body) may be more susceptible to men. Children and elderly are generally are more susceptible to chemical hazards and certain races may be genetically more vulnerable to certain chemical exposure. Increasing levels of exposure to or dose of a chemical will generally lead to more severe effects. For example, increasing exposure to benzidine dyes will result in a higher incidence of bladder cancers among the exposed population. This also means that at low levels of exposure to a chemical, the severity of the effect and the response decrease, and that at certain point there is no effect on health.

Acute Effects

Acute means 'of rapid onset and short duration'. A chemical having immediate effect on human body on brief/short exposure is categorized as 'acute effect'. Acute effects can result in a chronic disease. Relevant example is permanent brain damage resulting from acute exposure to trialkyl tin compounds or from severe carbon monoxide poisoning.

Chronic Effects

Chronic means 'of slow onset and long duration'. It usually refers to repeated exposure with a long delay between the first exposure and the appearance of adverse health effects. The delay between the time of the exposure to the onset of the health effect is called the latency period.

Systemic Effects

The effect of a chemical on the organs and fluids of the body after absorption and transport from the point of entry. Anemia (a deficiency of hemoglobin due to lack of red blood cells) is a typical systemic effect. It can be caused by a number of chemicals, e.g. lead, beryllium, benzene, cadmium and mercury compounds.

Cumulative and Latent Effects

The effect of some toxic substances like lead and cadmium is 'cumulative'. They continue to be build up in the body over time before they reach a 'critical level' which causes poisoning. The more serious health problem associated with hazardous chemicals is 'chronic and latent effects'. Symptoms of effects may arise 15 to 20 years after exposure such as in the case of 'asbestos poisoning'. These include 'birth defects', 'genetic' and 'neurological' disorders.

Table 1. Toxicity Levels of Some Hazardous Wastes and Chemicals

Contaminants	Maximum Concentrations (in ppm)
1. Arsenic	5.0
2. Benzene	0.5
3. Cadmium	1.0
4. Carbon tetrachloride	0.5
5. Chlordane	0.03
6. Chromium	5.0
7. 1,2, Dichloroethane	0.5
8. Endrin	0.02
9. Heptachlor	0.008
10. Lead	5.0
11. Mercury	0.2
12. Selenium	1.0
13. Vinyl chloride	0.2
14. Toxaphene	0.5

Health Hazards Associated with Persistent and Non-persistent Toxic Wastes and Chemicals

Chemical Compounds	Health Hazards
Persistent Organic Wastes and Chemicals High-molecular weight chlorinated and aromatic hydrocarbons, some pesticides (chlorinated insecticides like DDT, DDE, hexachlorbenzene, indane); PCBs, phthalates	Immediate toxic effects (acute and sub-acute) may occur at the source or point of release. Long term chronic toxicity may result. Transport of organic wastes from source can result in wide-spread contamination and bioconcentration in the human food chain.
Non-persistent Organic Wastes and Chemicals Oil, low molecular-weight solvents, some biodegradable pesticides (organophosphates, carbamates, triazines, anilines, ureas), waste oils and most detergents.	Toxicity to biota at the source or point of release and passing into human food chain. Acute to sub-acute toxic effects occur rapidly after exposure.

Chapter 8

CHEMICALS IN THE BREATHING AIR: THEIR POTENTIAL HEALTH RISKS AND IMPACTS

Several studies have confirmed that polluted air is responsible for increase in cancer by a third, infant mortality by two-thirds, heart ailments by 40 % and miscarriages by half since 1970. According to a study coal power plants in the U.S. cuts short nearly 24,000 lives, including 2,800 from lung cancer as well as nearly 38,200 from heart attacks each year. Each of those people whose lives were cut short lost an average of 14 years (WHO Reports, 1992-2004). From New York to California, air pollution takes a heavy toll on the health of the poor. Federal Center for Disease Control and Prevention in the U.S. places health cost to combat air pollution–related diseases in the U.S. at US $ 14 billion a year. (UNEP Report, 2001). UNEP estimates that over 3 million people in world die every year from air pollution related diseases, most of them poor women and children living in the urban environment. Of them about 2 million die from indoor pollution.

In Katowice city of Poland 15 % higher incidence of circulatory problem, 47 % higher incidence of respiratory problems, and 30 % more cancers are attributable to air pollution. Studies indicate that polluted air in Bratislava in Czechoslovakia is responsible for increase in cancer by a third, infant mortality by two-thirds, heart ailments by 40 % and miscarriages by half since 1970. In Hungary every 17th death and every 24th disability is directly or indirectly due to air pollution (UNEP Reports 1972-1992).

AIR POLLUTION AGGRAVATE FREE RADICALS FORMATION IN HUMAN BODY DAMAGING THE HEALTHY CELLS

Air pollution is aggravating the formation of 'free radicals' in human body. They are highly reactive molecules with unpaired electrons that float freely through the body seeking the 'healthy cells' to steal electrons and re-balance themselves. These causes damage to the cells they come in contact with and are linked not only to the degenerative diseases like cancer and arthritis, but the aging process itself.

HUMAN DISASTERS CAUSED BY AIR POLLUTION

There are several episodes of human disasters caused by air pollution in world. In 1952, chemical smog lasting several weeks led to the death of more than 4000 people in London City by acute aggravation of pre-existing respiratory problems. Earlier in 1872 and 1880 smog had killed 500 and 2000 people in London. In southern California thousand of people suffered from respiratory and heart problems due to smog. More and more people in automobile packed mega-cities are suffering from air toxicity related ailments like sore throat, clogging of sinuses, running nose, watering eyes, asthma, bronchitis, tuberculosis, loss of appetite and fatigue. Higher incidence of bronchitis is reported to be in the children in U.K.

Table 2. Air Toxics (Pollutants) Believed Dangerous to Human Health

Criteria Air Pollutants	Hazardous Air Pollutants (HAPs)
1. Sulfur Oxides	1. Asbestos
2. Carbon Monoxide	2. Benzene
3. Ozone	3. Mercury
4. Nitrogen Dioxide	4. Beryllium
5. Lead	5. Vinyl Chloride
6. Fine Particulate Matter	6. Inorganic Arsenic
	7. Radionuclides

Source: Air Pollution Control Engineering; McGraw-Hill (2000).

Table 3. Health Threats due to Some Hazardous Chemicals (Pollutants) in Environment

Pollutants	Source	Health Concern
1. Asbestos	Mines and building materials	Fatal mesothelioma and lung cancer
2. Benzene	Solvents used in Industries	Hepatotoxic
3. Arsenic	Copper smelters and cigarette smoke	Lung cancer
4. Mercury	Mining and material industries	Brain damage and nervous disorders, bowel
5. Vinyl chloride	PVC Industries	Lung and liver cancer
6. Tropospheric Ozone	Auto exhaust	Lung damage and impaired function
7. Airborne particulates	Auto exhaust	Eye and throat irritation
8. Carbon monoxide	Auto exhaust	Asphyxia, heart, vessels and brain damage
9. Sulfur dioxide	Coal powered plant and oil refineries	Respiratory tract damage
10. Nitrogen dioxide	Do + Auto exhaust	Respiratory illness and lung damage
11. Lead	Smelters, panel beating, car batteries	Brain damage in children, anemia
12. PAH	Diesel exhaust and cigarette smoke	Lung cancer

Source: U.S. Environmental Protection Agency.

CARCINOGENIC POTENCY OF CHEMICALS IN AIR

Many compounds known to occur as urban air pollutants are known to have carcinogenic potency. Indeed only about 10 % of the more than 2800 compounds that have been identified in the air have been assayed for carcinogenic potency.

UNEP reports that non-cigarette smoker lung cancer rate due to outdoor air pollution in world today is estimated to about 2000 cases per year as compared to the smokers which is 100,000 cases every year. By and large, the carcinogenic potency of air pollution resides in the particulate fraction. Polycyclic organic chemicals, along with a group of lower-boiling organics sometimes referred to as 'semi-volatiles' (including nitroaromatics), are associated with the particulate fraction and could have a prolonged residence time in sensitive sites in the respiratory tracts when inhaled.

FACTORS AFFECTING HUMAN HEALTH IMPACTS OF CHEMICALS IN THE AIR POLLUTANTS

Health impact of chemicals in the air depends upon several factors such as-

Individual's Duration of Exposure

The health effects of air pollution can be both short-term and long-term. Current interest in air pollution impact on human health is mostly directed at long-term, low concentration exposures which lead to 'chronic effects'. Short-term, high concentration exposures which lead to 'acute effects' occur only during industrial accidents and leakage of toxic gases into the atmosphere, the one which occurred during the Bhopal Gas Tragedy in India in 1984.

a). Short-Term Exposure

Short-term exposure to air pollutants can lead to headache, nausea, fatigue and irritation to eyes, sore throat, clogging of sinuses and running nose, loss of appetite and digestion related problems, upper respiratory and lung related infections like asthma and bronchitis, hypertension and skin problems.

b). Long-Term Exposure

Long-term effects can lead to chronic respiratory diseases, damage to the vital organs like brain, nerves, liver and kidneys; diabetes, heart disease and lung cancer specially to those who are habitual smokers. These health problems are on the rise in cities all over the world, and this is attributed directly or indirectly due to the air pollution. Urban people in both developed and developing countries suffer from several mental and physical health problems today. Evidences are gathering which suggest that the 'declining male sperm' count in men all over the world is directly linked with long-term exposure to air pollution.

Health Status of the Individual Exposed

People with heart conditions like angina or lung conditions like asthma or emphysema, the elderly people and children are naturally more susceptible to air pollution. Children are in fact, the worst sufferers of air pollution. According to UNEP (1996) more and more children in the cities of U.K. and U.S. are suffering from asthma directly attributable to air pollution. In the U.S. air pollution takes a heavy toll on the health of the poor. Those concentrated in urban areas are particularly badly affected. Over 57 % of whites, 65 % of African Americans, and 80 % of Hispanics live in 437 counties with substandard air quality. (UNEP Report, 2001). Numerous studies have indicated that poor (with nutritional deficiency) in the U.S. face greater health and environmental risks than the society at large.

Concentration of Chemical Pollutants in the Ambient Air

Greater the concentration of the air pollutants, higher is the risk to human health upon exposure. Air-pollutants are more concentrated in the areas of low air-current such as in the underground car parks and in motor garages. Wind can not only blow out pollutants but also dilute it. In modern homes with attached motor garage the indoor air can have greater concentrations of pollutants especially during the morning hours. Vehicles emit more pollutant during the cold start as the catalytic converters takes time to work.

Synergism

Whether the air-pollutants are acting separately or jointly with other toxic pollutants. This is called synergisms. The effect of two pollutants together is greater than the sum of the separate effects of the two. Sulfur oxides with particulates have more severe impact than SO_x and particles alone. A mixture of sulfur dioxide and sulfur trioxide has been shown to have more severe effects on the functioning of respiratory passages than the mathematical sum of their individual effects on the respiratory system. The same is true for the combined effect of exposure to more than one chemical at one time, or to a chemical in combination with other health hazards such as heat, noise or radiation.

Chapter 9

SOME KEY CHEMICALS IN THE AIR POLLUTANTS AND HEALTH IMPACTS

Air pollution from transport and industry contributes to an increased risk of death from cardiopulmonary causes, increased respiratory diseases, increased incidence of lung cancer in people with long term exposure, and adverse outcomes in pregnancy, such as premature birth and low birth weight.

HEALTH IMPACTS OF PARTICULATE POLLUTANTS IN AIR

Evidences suggest that small particulates are associated with an increased mortality risk. Health impacts may be due to the more toxic chemical pollutants adsorbed on the particle pollutants.

Researches done at University of California, U.S. and the Queensland University of Technology in Brisbane, Australia have found tiny pollutants (ultra-fine micro-particles) which are less than 0.1 micrometers in diameter and are linked to lung and heart diseases. They can penetrate deeper into the human lungs than the larger particles and carry most of the toxins, trace elements and the carcinogens. They are deposited with high probability in the lower parts of the human respiratory system and thus have the biggest health impact. They are so tiny that they can pass through the lungs directly into the blood stream.

Coarse particles, on the other hand, are often of natural origin (dust, pollen, seaspray etc.). They are deposited in the upper respiratory tracts and their health impacts are modest. Smaller particles (less than 2.5 microns) generally come from fossil fuels, especially by diesel combustion. Exhaust from diesel engines contains finer particulates than gasoline and these particulates contain PAHs (polyaromatic hydrocarbons) which are potent carcinogens and mutagens. WHO (2000) guidelines have been revised to reflect that there is 'no safe levels' of particulates. They have negative health impacts on humans no matter how low the concentration in the atmosphere. (GEO, 2006).

Researches done at Edinburgh University reveal that particulate matters produce high levels of charged particles called 'free radicals'. They can alter cell function and damage body's tissues. They have been found to stimulate a change in the blood, which make 'blood clots' more likely to form, which can lead to serious consequences like stroke and heart attack. Thus air pollution can be one of the several factors for increase in the heart and circulatory problems in the cities.

HEALTH IMPACTS OF SULFUR OXIDES IN AIR

Sulfur oxides (SO_x) are strong respiratory irritants that can cause considerable health damage at high concentrations. When inhaled, sulfur dioxide effects the lining of respiratory tracts and lungs and increase susceptibility to viral and bacterial infection. It can stimulate bronchoconstriction and mucus secretion in humans. Adverse effects were noted in humans at concentrations as low as 1 ppm of SO_2. The penetration of sulfur dioxide to the lungs is greater during mouth breathing than it is during nose breathing. An increase in the polluted airflow rate such as during exercise also markedly increase penetration to the deeper lung. Once deposited, sulfur dioxide dissolves in airways and in lungs fluid as sulfite or bi-sulfite and is readily distributed throughout the body. However, some residual sulfur dioxide persists in the respiratory system for a week or more after exposure.

HEALTH IMPACTS OF NITROGEN OXIDES AND ITS REACTION PRODUCT OZONE IN AIR

In the atmosphere and in the industrial devices NO reacts with O_2 to form nitrogen dioxide (NO_2) a brown gas which at high concentrations is a severe deep lung respiratory irritant that can produce pulmonary odema if inhaled at higher concentrations. It is deposited along the respiratory tree, mostly in the distal lung damaging the terminal bronchioles. It is also an important indoor pollutant. Another important concern about the NO_x is that they react with the hydrocarbons (Volatile Organic Carbons) and lead to the formation of ozone (O_3) in the troposphere (air at ground level) which is strong respiratory irritant and one of the principal constituents of urban summer eye- and nose-irritating smog.

NO + HC (Hydrocarbon) + O_2 + sunlight → NO_2 + O_3 (Ozone)

Ozone increases human susceptibility to infection and respiratory diseases and reduce lung function. It is also linked with rising 'asthma attack', allergic and cardio-respiratory disorders and death. In U.K. and the U.S. childhood asthma increased 5 times from 1979 to 1989, all due to the increasing amount of ozone in the breathing air. In the U.S., asthma accounts for 10 million missed school days, 1.2 million emergency room visits by children, and 500,000 hospitalization each year. (UNEP Report, 2001).

HEALTH IMPACTS OF CARBON MONOXIDE (CO) IN AIR

Carbon monoxide is different from most other air pollutants in its acute health effects. Most other air pollutants (sulfur oxides and nitrogen oxides) rarely cause fatalities due to short-term (acute) exposures, but their effects are much less likely to be reversible. However, the effect of CO on human health is practically reversible. Few hundred fatalities occur in the U.S every year due to exposure to high concentration of CO, mostly inside buildings with improperly vented heating systems, in idling parked cars with faulty exhaust system, and also in several industries. CO poisoning is the major cause of death in residential fires (about 4000 death every year) and coal mine fires (about 10 every year) in the U.S. (Nevers, 2000).

CO impairs the oxygen binding capacity of the blood hemoglobin and stops transport of oxygen to lungs causing 'aphyxia' and even death. CO binds with the hemoglobin in our blood, forming 'carboxy-hemoglobin' (COHb). Hemoglobin has roughly 220 times more affinity to bind with CO than with O_2.

Table 4. Health Impact (Poisoning Effect) of Carbon Monoxide

% of Hb Bound by CO	Health Impact
2.5 – 3.0	Cardiac function decrements in impaired individuals; Changes in RBC concentration.
4.0 – 6.0	Visual impairments; reduced work capacity
3.0 – 8.0	Routine values in smokers; smoker develop more RBC to compensate for loss of Hb
10.0 – 20.0	Slight headache; abnormal vision
20.0 – 30.0	Severe headache and feeling of nausea
30.0 – 40.0	Weak muscles; nausea; vomiting and irritability
50.0 – 60.0	Fainting; convulsions and coma
60.0 – 70.0	Coma; depressed cardiac activity and respiration
> 70.0	Fatal

Source: Air Pollution Control Engineering; McGraw-Hill (2000).

Hence even very small amounts of CO in the breathing air can cause significant amounts of our blood hemoglobin to be tied up with CO and unable to bind with vital oxygen (O_2) to transport it to our lungs and body tissues and make them starve for oxygen. Thus CO immobilize the hemoglobin for oxygen transport and kills the vital tissues. If over 70 % of the blood hemoglobin in the body is bound by CO as COHb and immobilized, it may prove fatal.

Smoking has similar effect on blood hemoglobin. Low birth weight and increased infant mortality in smoking mothers may be due to reduced oxygen supply to the fetus caused by the CO inhaled with the cigarette smoke. By reducing oxygen supply to vital organs CO can damage heart and blood vessels and also brain.

Chapter 10

CHEMICALS IN THE HUMAN ENVIRONMENT DUE TO VEHICULAR POLLUTION

In mega-cities urban inhabitants inhale vehicular pollutants on an average which is equivalent to smoking 10 cigarettes a day. Diesel vehicles emit over 100 micro-particles smaller than one micrometer in diameter, which are easily carried away by air currents and settles in the lower respiratory tracts of human lungs when inhaled. These particulates contain hundreds of organic compounds, several of which are carcinogenic. In Mexico City thousands people die every year due to vehicular pollution related diseases. Babies were reported to be born without brain (anenocephaly). They were found to have dangerous lead (Pb) levels in their umbilical cords.

Recent researches done at Los Angeles in the U.S. has indicated that people who live or work 50 meters or less downwind from a major highway or busy intersections could be exposed to up to 30 times higher than normal levels of potentially dangerous air pollutants. Study carried out in Germany indicate that the risk of heart attack was greatest within an hour of being in traffic. Research also shows that people in cars and buses in cities are exposed to 10 times the pollutants and chemical toxics that as on the side walk, largely because the emissions from the tailpipes of cars in front of them. People who live near major roads are at greater risk of dying from lung and heart problem. Air pollution worsens hardening of the arteries, blood vessels and disturb the heart's natural rhythm. Particulates increase inflammation, which can cause the rupture of plaques in arteries around the heart. Those ruptures can lead to blood clots that choke off blood supply and cause heart attack.

Chapter 11

CHEMICALS IN THE INDOOR AIR: THE HIDDEN DANGER IN MODERN HOMES

Deteriorating quality of the indoor breathing air by in-house air pollutants is a major concern today as people spend nearly 80 to 90 % of their time in their house or other indoor premises. Studies indicate that pollution levels in the homes are up to 20 times worse than the air outside. Some modern building materials give off 'radon' and 'asbestos' particles and volatile organic compounds (VOCs). Air-conditioning systems if not cleaned regularly can become a source of microbiological pollution. Smoking can be a big source of indoor pollution. Sick building syndromes, chronic fatigue, increasing rate of asthma, as well as declining sperm counts in men, have all been linked to air quality problems mainly indoor.

Indoor air pollution from biomass smoke have been found to be one of the four most critical environmental problems in developing countries. Some 3.5 billion people, mostly in rural areas, are exposed to high levels of indoor air pollutants from biomass smoke. Burning solid household fuels accounts for some 2.5 million premature deaths every year – about 6-7 % of the global disease burden, considerably more than due to ambient urban air pollution. The poorer a country, the more its people generally have to depend on traditional sources of energy like biomass (fuelwood and dung cakes), and the more likely they are to suffer from biomass smoke and die.

Noxious components like respirable particulates, carbon monoxide, nitrogen oxides, formaldehyde and polyaromatic hydrocarbons (PAH) such as benzo(a)pyrene found in the biomass smoke have been linked with the heart and lung diseases, cancer of lungs, nasopharynx, larynx and acute respiratory infections and middle ear infection in children. Low birth weights and perinatal mortality is caused if pregnant mothers are exposed to wood smoke for long hours. Benzo(a)pyrene is a known carcinogen and is also linked with depression in immune system responses.

High exposure can damage the respiratory system, eyes and immune system responses and make people susceptible to infection and diseases like tuberculosis, chronic obstructive pulmonary disease corpulmonale and lung cancer – and is associated with asthma, blindness, anemia. Extended exposure to high levels of biomass smoke can impair the clearing ability of the lungs and render them more susceptible to infection.

Recent researches indicate that people living in households relying more on biofuels for cooking are 2 to 3 times more likely to have active tuberculosis. Smoke can increase the risk of tuberculosis by reducing resistance to initial infection. It can promote active tuberculosis in people who are already infected. There are evidences that smoke can also cause cataract and aggravate trachoma and conjunctivitis which can result into blindness. Some 18 % of the blindness in India is attributed to the use of biomass smoke. Smoke is also more likely to aggravate asthma, triggering an attack rather than causing it. (UNEP Report, 2001).

Studies in India and South America have shown that exposure to indoor air pollution severely reduces lung function in children. Exposure to biomass smoke increases the risk of 'acute respiratory infection' (ARI) in children under 5 years of age. This is the single most important cause of morbidity and mortality worldwide, killing more than 3 million children under 5 every year and accounting for an estimated 9 % of the entire global disease burden. Children in Gambia riding on their mother's back as they cooked over smoky stoves were 6 times more likely to develop acute respiratory infection (ARI) than the unexposed children. (UNEP Report, 2001). Exposure to high indoor smoke levels is also associated with pregnancy-related problems such as stillbirths and low-birth weight.

Studies in India, Colombia, Mexico, Nepal and Papua New Guinea show that non-smoking women who have cooked on biomass stoves for many years exhibit a higher prevalence of chronic lung disease (asthma and chronic bronchitis) than those who have not.

Chapter 12

HEALTH IMPACTS OF CHEMICALS IN THE WATER

HEALTH RISKS OF NITRATES IN HUMAN FOOD AND WATER

Nitrate (NO_3)-nitrogen occurs naturally in food and water, usually at concentrations far below a level of concern for human health. Bacteria in the soil / water convert various forms of nitrogen to nitrate. In the recent years concern for nitrates in food and water has been growing alarmingly due to widespread use of chemical fertilizer (Urea) in agriculture. Nitrate is a potential health hazard. WHO recommends 10 mg/L of nitrate-nitrogen in water as the safe limit.

Friendly bacteria in our digestive system transform 'nitrates' to 'nitrites'. The nitrites oxidizes iron (Fe) in the blood hemoglobin to form 'methemoglobin' which then looses the capacity to absorb and transport sufficient oxygen to individual body cells. The disease has been called 'methemoglobinemia' or the 'blue- baby syndrome' and the infants under 6 months of age are more susceptible to this condition. Most humans over one year of age have the ability to rapidly reconvert the methemoglobin back to oxyhemoglobin.

a). Gastric Problems and Birth Defects

High nitrates may also cause hypertension in children, gastric cancers in adults, and fetal malformation in mother's womb. There are reports of potential 'birth defects' in pregnant women who drank water with high nitrates. Older youth and adults can tolerate higher levels of nitrate-nitrogen with little adverse affects. Older persons who suffers from gastrointestinal system disorder producing a pH level which allows for increased bacterial growth in the intestine may be at greater risk than the general population.

b). Cancer Risk

Potential cancer risk from high nitrates in water also exists. Nitrates can react with amines or amides in the human body to form 'nitrosamine' which is a known carcinogen and a mutagen (causing birth defects). Nitrates with pesticide residues in food may also form 'nitrosamines'.

HEALTH RISKS OF FLUORIDE IN WATER

The WHO permissible limits of fluoride in water is 0.7 to 1.2 ppm above which it may have toxic effects. The organs primarily affected by fluoride are bones and teeth. It impairs calcium (Ca) metabolism and pituitary water balance, cause dental and skeletal fluorosis (disintegration of bones and teeth), and even interfere with 'endocrine function'. It can causes 'crippling diseases' due to softening of bones in human beings. Fluorides also hardened the arteries, and cause cardiovascular diseases. Long term ingestion of water containing 8 ppm or more fluoride cause 'skeltal osteosclerosis' of vertebral column, pelvic girdle and ribs. Above 50 ppm it induces thyroid changes, above 100 ppm growth retardation, and above 125 ppm kidney damage. More than 5 gms can be fatal. (WHO Report).

Large number of cases of 'fluorosis' has been reported from India particularly the state of Rajasthan. Groundwater in some areas in Rajasthan is badly contaminated with high levels of fluoride and people suffer from various crippling diseases. Several works done by me at Indira Gandhi Centre for Human Ecology, Population and Environment Studies, University of Rajasthan, Jaipur, (1992-98) confirm these reports.

ARSENICOSIS DUE TO ARSENIC (AS) IN WATER

Arsenic (As) is widely distributed throughout the earth crust and may occur naturally in groundwater and surface water due to 'leaching' from the minerals in host rocks. Arsenic is also carcinogen and a mutagen (causing birth defects). Long-term exposure via drinking water can cause cancer of the skin, lungs, urinary bladder, and kidneys. The WHO established 10 ppb (parts per billion) as a provisional guideline for safe limit of arsenic in water in 1993. Arsenic pollution of groundwater was first noticed in the early 1980s in Bangladesh and West Bengal (India) where people suffered from arsenic poisoning (arsenicosis) by drinking water from hand-pump driven tube-wells.

DISRUPTION IN ENDOCRINE FUNCTION DUE TO CHEMICALS IN WATER

A wide array of industrial chemicals (some 45) specially the synthetic chemicals and the POPs in water sources can disrupt the genetically based messages through disruption of 'endocrine function' causing reproductive and developmental abnormalities, neurological and immunological problems and cancer. They have been called 'endocrine disrupting chemicals' (EDCs). Arsenic, dioxin, dibutylphthalate, dioctylphthalate, bisphenol-A and 17 *b*-estradiol are few of them. Arsenic, dioxin and bisphenol A are endocrine disruptors in very low-parts-per billion. They pose an even greater danger to the developing foetus in the womb. Because hormones are crucial in early development, endocrine disrupting chemicals may alter development of child's body and brain, and undermine the ability to learn, to fight off diseases and to reproduce. Increase in testicular, prostrate, and breast cancers have been blamed on endocrine disrupting chemicals.

CHLORINATED WATER AND HEALTH HAZARD

Evidences are gathering that disinfecting public water supplies with 'chlorine' which is a common practice all over the world, might actually lead to the formation of 'volatile halogenated derivatives', some of which are known carcinogens such as the 'trihalomethanes' (THM).

Chapter 13

HEALTH IMPACT OF CHEMICAL CONTAMINATION OF FOOD AND FARM SOIL DUE TO WIDESPREAD USE OF AGROCHEMICALS

Greater reliance on chemical pesticides has posed considerable risk to human health and the environment. Food contamination due to ambitious use of chemical pesticides (to boost productivity) is a growing problem all over the world but more in developing countries. Organic chemicals and heavy metals that persist in the soil such as cadmium (Cd) are absorbed by the food crops and pass into the human food chain. They pose serious risk of adverse effects on human health, even leading to cancer, reduced fertility and neurological damage. Most pesticides are carcinogenic and neurotoxic. In the U.S. some 45,000 humans are poisoned to some degree by pesticides every year. U.S. scientists predict that up to 20,000 Americans may die of cancer, each year, due to the low levels of 'residual pesticides' in the chemically grown food. (UNEP Report, 2002).

Chapter 14

SYNTHETIC CHEMICALS USED IN HUMAN DEVELOPMENTAL ACTIVITIES AND THEIR POTENTIAL HEALTH RISKS AND IMPACTS

Human ingenuity and technology has developed several synthetic chemicals to be used in diverse developmental activities. Some of the more well know chemicals, known more for their nuisance value as potentially toxic, carcinogenic (inducing cancer) and mutagenic (causing birth defects) substance are described here. A number of them have now been banned from production in several industrialized countries. As the philosophy of Cleaner Production is embraced and adopted by nations more and more such hazardous chemicals will be banned. But some of them (persistent chemicals) with long residency time will still remain in the human environment to haund us.

THE HAZARDOUS ORGANIC CHEMICALS USED IN HUMAN DEVELOPMENT AND RISK TO HUMAN HEALTH

1). Acetonitrile

Acetonitrile, a byproduct of acrylonitrile manufacture which is widely used as an extractive distillation solvent in the petrochemical industry and as a solvent for polymer spinning and casting in textile industries. Studies of kinetics and metabolism indicate that acetonitrile is readily absorbed by all routes and rapidly distributed throughout the body, where it is converted to 'cyanide'.

Human Health Risks: Acetonitrile has low toxicity as such, because of its rapid volatilization and biodegradability, but once converted into cyanide it induces toxic effects similar to those like cyanide poisoning, with prostration followed by seizures. (UNEP and WHO Reports, 1992-2006).

2). Benzene

Benzene is a colorless solvent used as an additive in petrol, and also manufactured in extremely large quantities worldwide for their use in plastics, dyes, detergents and insecticides. It is also a naturally occurring chemical found in crude petroleum.

Human Health Risks: Benzene is a well established carcinogen. WHO finding has established that benzene damage 'bone marrow' and is a confirmed human 'leukaemogen'. A short period of exposure benzene can cause immune system depression, temporary nervous disorders and anemia. Benzene can bound to DNA and cause mutation. It thus poses serious health risks to both the general human population as well as the industrial workers. (UNEP and WHO Reports, 1992-2006).

3). Bisphenol A

It is used in the manufacture of polycarbamate and other plastics, in the lining in tin cans, floorings, enamels and varnishes, adhesives and nail polish, compact disc, electric and electrical appliances.

Human Health Risks: The chemical has been found in the blood of pregnant women, in the umbilical chord of the newborn at birth and in placental tissues, at levels where it can alter development. It has been implicated with spontaneous miscarriages and birth defects, including Down's Syndrome (UNEP Report, 2004).

4). Chlorobenzene

Chlorobenzenes (mono-chlorobenzenes, di-chlorobenzenes tri-chlorobenzenes tetra-chlorobenzenes and penta-chlorobenzenes) are produced in huge quantities for use as solvents and chemical intermediates in the production of pesticides, phenols, DDT, di-isocyanates, aniline and in the manufacture of wide range of consumer and commercial products e.g. moth repellents, air fresheners, paints, adhesives, polishes, waxes, pharmaceuticals and rubbers. Other application include use as fiber swelling agent and dye carrier in textile processing, as a tar and grease remover in cleaning and degreasing operations, as solvent is surface coating and surface coating removers; and as heat transfer medium. It can enter into the atmosphere as fugitive emissions from the pesticide industries and from industries using them as solvent. It also enters the environment via pesticide application, industrial effluents and leachate in water bodies. Volatilization of chlorobenzene from dry contaminated soil is also very high. Incineration may lead to emission of polychlorinated dibenzo-*p*-dioxins and dibenzofurans.

Human Health Risks: Chlorobenzenes have been found to affect central nervous system, cause irritation of eyes and upper respiratory tract.(UNEP and WHO Reports, 1992-2006).

5). Cresols

Cresols have a wide variety of uses in developmental activities- as solvents and disinfectants and as chemical intermediates for pharmaceuticals, fragrances, antioxidants, dyes, pesticides, and resins. It is also used in the production of soaps, lubricating oils, motor fuels and rubber polymers and in the manufacture of explosives.

Human Health Risks: Cresols are released in the air and water during production and use and general population may be exposed from cresols in air, drinking-water, food and beverages, and consumer products, such as soaps and disinfectants. Studies of acute poisoning in workers indicate that it may cause abdominal pain and vomiting, haematological changes, kidney failures, coma and death. Dermal contact can cause severe burns and scarring of the skin.

6). Dichloro Diphenyl Trichloroethane (DDT)

The chemical pesticide DDT came as a boon when it was first synthesized in the 1950's to eradicate mosquito causing 'malaria' (a curse for the people in the tropical countries) and the crop pests, and helped reap the green revolution. Soon it turned out to be a bane when Rachel Carson in her book '*Silent Spring*'(1962) indicated about its ill effects on the human health and the environment. Although DDT was prohibited for use within the U.S. as early as in 1972, the U.S. still manufacture over 18 million kg a year for export, largely to the Third World developing nations. (Myers, 1994). Wholesale use of DDT has polluted all major rivers of world and in absorption of tiny amounts by small fishes. They are 'persistent pollutant' and can remain in the environment for more than 20 years. They are found almost everywhere- in our food, soil, air, and water. Once in the environment DDT undergo a process of 'bio-amplification' by 10 – 100 times in the food chain and as a result their eventual concentration increases by million times, by the time they reach the human body. DDT production has however, now been banned in several countries.

Significant amount of DDT from the soil and water has 'bioaccumulated' in the cereals and pulses and the meat and milk consumed by the human beings. Varying levels of DDT has been reported in the cereals, pulses, fruits, vegetables, milk and meat- 1.6-17.4 ppm in wheat, 0.8-16.4 ppm in rice, 2.9-16.9 ppm in pulses, upto 5.0 ppm in green vegetables, and as high as 68.5 ppm in potato. Highest amounts of DDT have been reported in food with high fat contents like meat and milk. In India human fat DDT residue ranged from 1.8 ppm in Lucknow to 22.4 ppm in Ahmedabad; HCH ranged from 1.6 ppm in Mumbai to 7 ppm in Bangalore. Another serious cause of concern is that lactating mother's milk all over the world contains significant amount of DDT. It means that toxic pesticides have already passed into the human body system. An average Delhite is reported to carry a load of more than 50 ppm of total organochlorines. (Rao, 1997)

Once in the environment DDT undergo a process of rapid 'bio-amplification' magnifying their effective dosage by 10 – 100 times at each trophic level in the food chain. As a result their eventual concentration can increase a million times by the time they reach the human body. Significant concentration of DDT ranging from 3 ppm to 26 ppm have been found in the body fat of human population across the world. Man is at the receiving end of the widespread pesticidal usage, which he had intended to use for controlling his enemy (the pests). We carry a heavy loads of 'persistent pesticides'. What is more alarming is that even the Eskimos living in the remote corner of world in the North Pole have significant amount of DDT in their body fat. Data below shows the situation in the 1960's. What could be the DDT amount of human population today after 40 years of widespread use ?

Table 5. DDT in the Body Fat of Some Human Population

Population	Year	Number of Samples Studies	Concentration of DDT (ppm)
1. Alaska (Eskimos)	1960	20	3.0
2. Canada	1959-60	62	4.9
3. England	1964	100	3.9
4. France	1961	10	5.2
5. W. Germany	1958-59	60	2.3
6. Hungary	1960	48	12.4
7. India (New Delhi)	1964	67	26.0
8. Israel	1963-64	254	19.2
9. U.S. (Kentucky)	1961-62	130	12.7
10. U.S. (All zones)	1964	64	7.6

Source: *Limits to Growth;* Report of the Club of Rome (1971).

Human Health Risks: The health threats of DDT is multiplied because it tends to lodge in the fatty tissues of living organisms and are not secreted. In laboratory animals DDT exposure was associated with an increased frequency of cancer. It may have the same impact on the *Homo sapiens* (humans) at higher concentrations. The most common finding in USA is an association between 'breast cancer' in women and higher levels (50-60 %) of DDT and its derivative DDE in their tissues. A study made by US Centers for Disease Control in 2001, reported a strong relationship between DDT contamination in mothers and the likelihood of 'pre-term birth' of their infants. US experienced an epidemic of pre-term birth during the hey-days of DDT use in the 1960s when the 'green revolution' was in its full swing in the farms. DDT may have caused up to 15 % of infant mortality in the US those days. (UNEP and WHO Reports, 1992-2006).

7). Dioxins and Furans

Dioxins and furans are a family of chemicals comprising 75 different types of compounds and 135 related compounds known as furans. They are undesirable 'byproduct' of some chemical processing in industries or generated during incineration of chlorinated hazardous wastes. They are unintentionally created whenever chlorine-based chemicals such as the vinyl chloride are produced, used or burnt. Polychlorinated dibenzo-*para*-dioxins

(PCDDs) are most dangerous. It is formed as inadvertent by-products, sometimes in combination with polychlorinated dibenzofurans (PCDFs), during the production of chlorophenols and chlorophenoxy herbicides. PCDDs and PCDFs may also be produced during metal-processing and in the bleaching of paper pulp with free chlorine. Manufacture of polyvinyl chloride (PVC) plastics is responsible for a greater share of the world's annual burden of dioxins in environment. Large amount of dioxins are produced during the various stages of PVC production and disposal by incineration. Abundance of PVC items in medical waste which is necessarily incinerated release large amount of dioxins.

Human Health Risks: Dioxins are persistent organic pollutant (POP) and are ubiquitous in soil, sediment, and air. Most human exposure to these toxic chemicals occurs from the consumption of meat, milk, eggs, fish etc. There is no 'safe limit' of dioxin exposure. Only a few parts per billion of the potentially toxic compound may be sufficient to cause a health hazard following a brief exposure. Of the PCDDs, 2,3,7,8- tetrachloradibenzo-*para*-dioxin (TCDD) is among the most toxic molecules ever known. It is implicated in weakening the 'human immune system' and impairment in 'fetal development'. Dioxins has been shown to cause severe reproductive deformities in mice and monkeys and its suspected impact is also on the humans (at levels 100 times lower than those associated with its cancer causing effects). Cancer, birth defects, endometriosis; hormonal, neurological, liver and kidney damage are attributed to dioxin. TCDD produces increased risks for all cancers combined, rather than for any specific site, indicating that it is an unprecedented multi-site carcinogen. After the Vietnam War in 1973, the Vietnamese children exposed to 'Agent Orange' (contains dioxin) suffered birth defects. The American soldiers using the chemical also suffered.

High levels of dioxins have been found in the breast milk of women in several parts of world. A new study has found that women exposed to high levels of one form of dioxin after the 1976 industrial disaster in Seveso, Italy have an increased risk of breast cancer. (UNEP Report, 2004).

8). Dichloroethane (DCE)

DCE is an industrial chemical mainly used in the manufacture of vinyl chloride (raw material in PVC plastic industries) and also in the production of various chlorinated solvents, as a fumigant, and in the production of anti-knock additives for petrol (gasoline) replacing the tetraethyl lead.

Human Health Risks: Study reveals that DCE is a probable carcinogen and has been shown to be 'genotoxic' in *in-vitro* and *in-vivo* assays. It is linked with pancreatic cancer and leukemia. (UNEP and WHO Reports, 1992-2006).

9). Diethylhexyl Phthalate

DEHP is produced in large quantities for use as a resin-softening plasticizer with major application in the production of polyvinyl chloride (PVC) used in the construction and packaging industries and in components of medical devices, including tubes used in blood and solute transfusions and dialysis.

Human Health Risks: Experimental studies indicate testicular atropy and neoplastic effects on the liver of rats and mice. Impact on human beings is being studied. Some evidences link it with affecting our behaviour. DEHP, however, rapidly photodegrades in atmosphere. (UNEP and WHO Reports, 1992-2002).

10). Dimethylformamide

It is a solvent widely used as an additive and as an intermediate in chemical industries and is produced in large quantities all over the world. Largest use is in the production of acrylic fibres and polyurethanes. It is also used in the manufacture of pharmaceutical products.

Human Health Risks: Studies indicate that dimethylformamide is a hepatotoxic agent. Both teratogenic and embryotoxic effects have been indicated. (UNEP and WHO Reports, 1992-2006).

11). Ethylene Oxide

It is an important raw material for producing major consumer goods in virtually all industrialized nations.

Human Health Risks: There is evidence of a link between exposure to ethylene oxide and the risk of lymphatic and haematopoietic cancer. (UNEP and WHO Reports, 1992-2006).

12). Ethylbenzene

The chemical is widely used in the production of styrene (one of the most important monomers used worldwide in plastic industry) and also xylene (used as solvent in paints and lacquers), and in the rubber and chemical manufacturing industries). It is emitted from vehicles and industries.

Human Health Risks: The chemical has low acute and chronic toxicity but is not carcinogenic or mutagenic. It also undergo photo-oxidation and biodegradation. However, the product styrene has been found to be carcinogenic. There is evidence of a link between exposure to styrene and the risk of lymphatic and haematopoietic cancer. (UNEP and WHO Reports, 1992-2006).

13). Formaldehyde

It is a gas produced industrially in large quantities for a wide range of applications, including the production of glue for particle boards and plywood and the manufacturing of sterilizing and disinfecting agents, medicines, cosmetics, and several consumer products. Hospitals and scientific facilities use it as sterilizing and preserving agent. There can be uncontrolled emission of formaldehyde from building materials and furniture thus contributing significantly to indoor pollution.

Human Health Risks: Studies have revealed possible cancer link to nasal and nasopharyngeal tumors from exposure to formaldehyde but not very high. However, it is not teratogenic or mutagenic causing birth defects. (UNEP and WHO Reports, 1992-2006).

14). Hydroquinone

Hydroquinone is a phenolic compound manufactured for a large variety of commercial applications, including use as a developer in black-and-white photography, in the production of medical and industrial X-ray films, in the manufacture of rubber antioxidants and antiozonants, and antioxidants for food preservation, and as a chemical intermediate for the production of agrochemicals and performance polymers. It is also used in cosmetics and skin ointment (cream) to treat disorders of pigmentation.

Human Health Risks: Recent studies indicate that co-exposure to hydroquinone and various other phenolic compounds can greatly potentiate the toxic effects of the individual compounds, causing cyto-toxic, immuno-toxic, and geno-toxic effects. (UNEP and WHO Reports, 1992-2006).

15). Hexachlorobenzene (HCB)

HCB has historically had many industrial and agricultural uses but due to its acute adverse affects on human health and environment several nations stopped production in the 1970s. However, inadvertent production of this highly persistent chemical continues in the form of by-products and impurities generated during the manufacture of chlorinated solvents, chlorinated aromatics, and chlorinated pesticides. HCB is widely dispersed in ambient air and in surface water but the concentration is low.

Human Health Risks: Studies indicate that HCB is carcinogenic in animals and undergoes significant bioaccumulation and biomagnification in the food chain. It has adverse non-neoplastic effects in animals, at relatively low doses, and the organs affected are liver, lungs, kidneys, thyroid, reproductive tissues, nervous and immune systems. Accidental poisoning incident by HCB occurred in Turkey in 1955-1959, when HCB-treated wheat grains was used to make flour and bread. More than 600 cases of enlarged liver, enlargement of the thyroid glands and lymph nodes and arthritis were reported. Nursing infants of exposed mothers developed a disorder called 'pink sore' and most died within a year. Follow-up of survivors at 20 and 30 years revealed persistent abnormalities. (WHO Reports 1992-2006).

16). Methylene Chloride

It is a highly volatile, stable and neuro-toxic chemical and is widely used in metal cleaning, in the chemical and pharmaceutical industries, and in the manufacture of polyurethane foam. It is also used in the consumer products such as the paint removers and in aerosols.

Human Health Risks: Study reveals that the chemical is rapidly absorbed through the lungs and the gastrointestinal tract and it can cause reversible depression of the central nervous system and also carboxyhaemoglobin formation. Studies also reported about liver and renal dysfunctions, haema tological effects, and neurophysiological and neurobehavioural disturbances. (UNEP and WHO Reports, 1992-2006).

17). Methanol

Methanol is extensively used as industrial solvent and is produced in huge amounts all over the world. It is also an intermediate chemical in the manufacture of methyl tertiary butyl ether (MTBE) used as additive in gasoline, formaldehyde, acetic acid, glycol ethers and a variety of consumers products. Consumer products containing methanol include paints, varnishes, shellacs, mixed solvents in duplicating machines, wind shield washer fluids, cleansing solutions, glues and adhesives, and in foods and soft-drinks containing the synthetic sweetener 'aspartame'.

Human Health Risks: Studies shows that exposure to methanol induces a wide range of concentration dependent teratogenic (causing genetical defects) and embryolethal effects (killing fetus). Methanol poisoning can lead to depression of central nervous system (CNS), blindness, coma and death. (UNEP and WHO Reports, 1992-2002).

18). Methyl Ethyl Ketone (MEK)

Because of its excellent properties as a solvent, MEK is widely used in the application of protective coatings and adhesives, in magnetic tape production, in the dewaxing of lubricating oil, and in food processing and textile dyeing industries. MEK is also a common ingredient in consumer products such as varnishes and glues. Environmental sources of MEK are volatilization from building materials, tobacco smoke and consumer products. MEK has been found in drinking water too.

Human Health Risks: Principal toxic effects of MEK stems from its well documented ability to potentiate the toxicity of two classes of organic solvents- the unbranched aliphatic hexacarbons and haloalkanes. Chronic co-exposure with these organic solvents represents a significant occupational hazard for industrial workers. Intentional abuse of solvents containing MEK can cause severe injuries, permanent disabling and even death. (UNEP and WHO Reports, 1992-2002).

19). Polychlorinated Biphenyls (PCBs)

PCBs are group of oily, colorless, volatile organic fluids used in several industrial activities such as in transformers and power capacitors, electrical insulators and as hydraulic fluids and fire retardant. They are added to paints, copying and printing paper inks, adhesives and plastics. There are over 200 types of PCBs. These chemicals, which are now ubiquitous in the environment, have been used commercially since 1930 but have now been banned in several countries including in Australia. PCBs are unusually toxic and persistent organic pollutant (POPs). They were originally released in the tropical and temperate environments long back.

Like DDT, PCBs are transported to long distances by the atmosphere and have even reached the poles in the Arctic and the Antarctica. The indigenous Inuit societies of northern Canada, Alaska and Greenland have very high levels of PCBs and DDT in their blood and lipid tissues. Once in the Arctic, they degrade very slowly and bioaccumulate. Once in the environment PCBs undergo a rapid process of 'bio-amplification' magnifying their effective dosage by 10 – 100 times at each trophic level in the food chain. As a result their eventual concentration can increase a million times by the time they reach the human body. Samples in the North Sea indicated PCB concentration of 0.000002 ppm in sea water which biomagnified to 160 ppm in marine animals. They bio-accumulate in the liver and the adipose tissues, transported through the placenta and distributed to milk.

Human Health Risks: Like DDT, PCBs are found almost everywhere- in our food, soil, air, and water and accumulate in fatty tissues in man and animals. Studies reveal that PCBs are immunosuppressive and act as tumor promoters. PCBs have been linked to birth defects, reproductive failures, liver problems, skin lesions, cancers and tumors. They are highly carcinogenic, inhibit cell growth and enzyme function; can cause asthma, hair loss, and joint deformity at birth. Children born with high levels of PCBs were found to have smaller head circumference, lower IQs, shorter attention spans and weaker reflexes. For the general population food items and for babies, the breast milk is most important sources of exposure.

Women exposed to PCBs around North America's Great Lakes have given birth to children with delayed motor development and dramatically lower intelligence. (UNEP and WHO Reports, 1992-2006).

20). Polybrominated Biphenyls (PBBs)

It is primarily used as 'flame retardants' for plastics in electronics industries. In computers it is mainly used in printed circuit boards, in connectors, in plastic covers and cables. It is also used in plastic covers of TV sets and in domestic kitchen appliances.

PBBs have been found in sediments of polluted lakes and rivers and in the Arctic seals and fishes of several regions indicating its wide persistence in the environment. It has been found to be 200 times more soluble in landfill leachate and have great risk of contaminating groundwater.

Human Health Risks: Researches in the U.S. found that exposure to PBBs may cause an increased risk of cancer of the digestive and lymph systems.

21). Polybrominated Diphenylethers (PBDEs)

Like PBBs it is also used as 'flame retardants' for plastics in electronics industries and in fabrics. They have been found in the Arctic wildlife in beluga whales where the levels have increased 6.5 times 1982 to 1997 (GEO Report, 2005/06). It has also been found in the human breast milk and its level is doubling every 5 years.

Human Health Risks: Several studies have indicated PBDE might act as 'endocrine disruptor'. Some studies have shown that the chemicals could be toxic to the immune systems and could affect the neurobehavioral development. Other studies have shown that PBDE, like many halogenated organics, reduces levels of the hormone thyroxin in exposed animals and have been shown to cross the blood brain barrier in the developing fetus.

22). Polycyclic Aromatic Hydrocarbons (PAHs)

PAHs are group of over 100 chemicals of which anthracene, benzo(a)pyrene, naphthalene, pyrene are common. They are formed by the incomplete combustion of coal (power plants), oil and wood. Napthalene, also known as 'mothballs', is used in making dyes, explosives, plastics, lubricants and moth repellents. Anthracene is used in dyes, insecticides and wood preservatives

Human Health Risks: PAH registers 1.3 on a health hazard spectrum. Benzo(a)pyrene is reported to be toxic and possibly mildly carcinogenic.

23). Phenol

Phenol is the basic chemical from which a number of commercially important materials are produced including phenolic resins, bisphenol-A, caprolactam, alkyl phenols, and chlorophenols. Environmental sources of phenol are the steel and textile industries. It is however biodegradable and unlikely to persist and bioaccumulate.

Human Health Risks: Phenol is highly neurotoxic chemical affecting the CNS. It can damage liver and kidney, affect respiratory systems and retard growth even at short-term exposure. Absorption of phenol through skin contact is very rapid and death results from collapse within 30 minutes to several hours. Studies are being made to evaluate its genotoxic and carcinogenic potentials. (UNEP and WHO Reports, 1992-2006).

24). Pentachlorophenol (PCP)

Once used in tremendous volumes as a biocide in leather tanning, wood preservation, the paper and cellulose industry and in paint industry has now been phased out and even banned from production.

Human Health Risks: Human poisoning by commercial PCP has occurred, usually associated with occupational exposure. The chemical is absorbed readily through the skin and has been detected in the urine. High-level exposure can result in elevated body temperature up to 42° C, profuse sweating and dehydration, marked loss of appetite and loss in body weight, tightness in the chest, generalized weakness, headache, nausea and vomiting, incoordination, early coma and even death. PCP acts at cellular level to uncouple oxidative phosphorylation during respiration, the target enzyme being Na^+, K^+-ATPase.

25). Styrene

It is one of the most important monomer used worldwide in plastic and polymer production.

Human Health Risks: Studies have indicated a definite link between exposure and the risk for lymphatic and haematopoietic cancer.

26). Trichloroethane (TCE)

TCE is a chlorinated hydrocarbon widely used in the cleaning and degreasing of metal and as a solvent in many industrial and consumer products. TCE has potential to deplete ozone and also induce global warming. The chemical is rapidly transported to the stratosphere and have long residency time. It can also rapidly leach into ground contaminating even the deep aquifers. No wonder then, that TCE is now found in several ground and surface water sources all over the world. (UNEP and WHO Reports, 1992-2006).

Human Health Risks: The abuse of this chemical has resulted in large number of fatalities all over the world. *It is an important chemical in the indoor air pollution due to its presence in several consumer products.* Both acute and long term inhalation exposure affect the central nervous system. Exposure may also cause damage to the vital organs heart and liver.

27). Vinyl Chloride

Vinyl Chloride is used as a raw material in PVC plastic manufacturing industries. PVC is the most widely used plastic material in developmental activities all over the world. It is used in everyday consumer electronics (computers, TVs, audio equipments etc), household items, pipes, upholstery etc. In fact this 'commonplace plastic' is one of the biggest contributors to the flood of toxic substances saturating the planet Earth and plaguing the human society. This poison plastic PVC contaminates the human environment throughout its lifecycle during its production, use and disposal. It may get dissolved in food, blood and medicine when these articles are contained in PVC bags. Large amount of dioxins are produced during the various stages of PVC production and disposal by incineration.

Human Health Risks: The chemical is highly carcinogenic and has been implicated with inducing liver cancer. A correlation between the occurrence of 'angiosarcomas' (liver cancer) and the workers handling vinyl chloride was found in the U.S. Polyvinyl chloride (PVC) destroys fertility of the animals and also affect respiratory systems. When consumed in soluble forms, PVC may cause paralysis, skin irritations and damage to bones. (UNEP and WHO Reports, 1992-2006).

28). Vinylidene Chloride

It is a vinyl chloride copolymers used mostly for the packaging of foods (commercial packing films and as household wraps), as tapes, as metal coatings in storage tanks, in building structures, as moulded filters, valves, and pipe fittings.

Human Health Risks: Main health hazards associated with VDC is the depressed functions of the CNS, skin and eye irritation. It may also be a possible carcinogen and a mutagen on long-term exposure but that awaits more studies. (UNEP and WHO Reports, 1992-2006).

29). Xylenes

It is an aromatic hydrocarbon blended into petrol and used in a variety of applications as solvents, mainly in the paint and printing industries. Xylenes in the human environment thus results from the automobile exhaust and from its use as a solvent. It is however, readily degraded in the environment through photo-oxidation. The isomer *ortho*-xylene is more persistent in the soil.

Human Health Risks: There is evidence of chronic effects on the CNS following exposure at moderate concentrations. It may have an acute impairing affect on the sensory-motor and information-processing functions of the CNS. No carcinogenic or mutagenic effects of xylenes have been found. Fortunately also, xylene is rapidly and efficiently metabolized, with more than 90 % biotransformed to methylhippuric acid and excreted in urine. (UNEP and WHO Reports, 1992-2002).

Chapter 15

HEAVY METALS USED IN DEVELOPMENTAL ACTIVITIES AND RISKS TO HUMAN HEALTH

Since 1950 there has been a tremendous increase in the use of 'heavy metals' to satisfy our demands for consumer goods. Annually, 0.5 million tones of cadmium, 310 million tones each of chromium and copper, 240 million tones of lead, 20 million tones each of mercury, manganese and nickel; and 250 million tones of zinc are mined and released into the human environment. Certain heavy metals are highly 'toxic'. Metals that provoke 'immune reactions' include mercury (Hg), gold (Au), platinum (Pt), beryllium (Be), chromium (Cr), and nickel (Ni). As a general rule, the ionic form of heavy metal is believed to be carcinogenic. A critical determinant of the metabolism and toxicity of heavy metals and other toxic compounds is its 'biological half-time' i.e. the time it takes for the human body to excrete half of the accumulated amount. The greater the BHT, the more severe will be the toxic effects. Persons at either end of the life span, whether they are young children or elderly people, are believed to be more susceptible to toxicity from exposure to a particular level of metal than most adults. Children have higher gastrointestinal absorption of metals, particularly lead (Pb). The rapid growth and rapid cell division that children's bodies experience represent opportunities for genotoxic effects. Background levels of chemical contamination can make children 'less resistant' to infectious agents. Lifestyle factors such as smoking or alcoholism may have influences on toxicity. Cigarette smoke by itself contains some toxic metals like cadmium (Cd).

1). ARSENIC (AS)

Arsenic is a product of several industrial activities- mostly mining for base metals (gold and tin), copper, zinc and lead smelters and coal power plants. Arsenic ore (arsenopyrite) is also specially mined, roasted and refined to produce arsenic oxide used in several developmental activities - in agrochemicals (as pesticides), as wood preservatives, in treatment of sulfide ores, hardening of copper, lead and other alloys, in glass and pharmaceutical industries and in the manufacture of solar photovoltaic cells (PVC). About 70 % of the world production of arsenic is used in timber treatment, 22 % in pesticides, and the remainder in glass, pharmaceuticals and metallic alloys.

Human Health Risks

Arsenic is highly toxic carcinogen (Category I) and also a mutagen / teratogen (harming fetus). It registers 2.3 on the health hazard spectrum (HHS), 3 being most hazardous and hence arsenic is rated high as human health hazard. Long-term oral exposure via drinking water or breathing air can cause cancer of the skin, lungs, urinary bladder, and kidneys. Other tumors that have been associated with arsenic exposure are 'hem-angiosarcoma' of the liver, lymphomas, leukemia, and nasopharyngeal. The time between initiation of exposure and occurrence of arsenic-associated lung cancer has been found to be on the order of 35 to 45 years.

Chronic exposure to inorganic arsenic compounds may lead to neurotoxicity of the peripheral and central nervous systems. A number of sulfydryl-containing protein and enzyme systems have been found to be altered by exposure to arsenic. Arsenic also effects mitochondrial enzymes and impairs tissue respiration. Ingestion of large doses (70-180 mg) of arsenic may be fatal. The major source of occupational exposure to arsenic is in the manufacture of pesticides, herbicides, and other agricultural products.

The biological half-life of ingested inorganic arsenic is about 10 hours, and 50 to 80 % is excreted in about 3 days. Arsenic has a predilection for skin and is excreted by desquamation of skin and in sweat. It also concentrates in nails and hairs. Arsenic in nails produce transverse white bands across fingernails and can work as a bio-indicator of arsenic poisoning. Anorexia, hepatomegaly, melanosis, and cardiac arrhythmia are some of the symptoms of arsenic poisoning. (UNEP and WHO Reports, 1992-2006).

2). ASBESTOS (AB)

Asbestos is a mineral fiber that has historically been used in industrial activities for insulating and fireproofing, as a wide variety of construction and industrial materials, and in equipments.

Human Health Risks: Asbestos is highly hazardous when inhaled as microscopic fibers. It has been found to be carcinogenic causing 'mesothelioma' (lung cancer) which is rapidly fatal. The main organs affected are pleura and the peritoneum of lungs. Even occasional handling of asbestos and slow inhalation of tiny asbestos fibers can cause lung cancer. The most serious thing is that it has a 'latent effect' i.e. cancer may develop in 20 to 30 years time.

3). BERYLLIUM (BE)

Beryllium, a brittle metal is extracted and produced in large quantities every year. It has major application and use in electronics and micro-electronics industries, in nuclear energy generation, x-ray windows, in production of military devices, including satellites, atomic bombs, missiles, and other weapons. It has also proved its superiority as a structural material for aircraft and spacecraft. Another major use is as an alloy.

Human Health Risks: WHO evaluation of impact of beryllium on human health has found the occurrence of both acute and chronic beryllium disease. Half of the inhaled beryllium is cleared in about two weeks; the rest is removed slowly, and a residuum becomes fixed in the tissues probably within fibrotic granulomos. The half-life of beryllium in tissues is relatively short, except in the lungs and is also excreted in urine. Acute pulmonary disease (*Chemical Pneumonitis*) from inhalation of beryllium is a fulminating inflammatory reaction of the entire respiratory tract. There is sufficient evidence to establish the role of beryllium in development of human 'lung cancer'. It has also been found to be highly 'teratogenic' and can provoke contact allergic reactions.

4). CADMIUM (CD)

Cadmium is mined especially for its industrial use in electroplating industry due to its non-corrosive properties. It is used as a cathode material for nickel-cadmium batteries. It is also used as a color pigment for paints and plastics and in steel production. Cadmium is also a by-product of zinc and lead mining and smelting industry, and is present in all food processing equipment, kitchenware enamels, pottery glazes and plastics.

Human Health Risks: Cadmium is regarded as the potentially dangerous industrial chemical. It is a 'cumulative poison' affecting calcium metabolism and is transported in blood by binding to red blood cells and the albumin in plasma. Cadmium concentrates in the liver, kidneys, pancreas and thyroids, damage kidneys and blood vessels. Highest cadmium concentrations (50 to 75 %) are generally found in the renal cortex and as exposure levels increase, a greater portion of the absorbed cadmium is stored in the liver.

Half-life of cadmium in human body is about 30 years. Acute exposure to cadmium develop flu-like symptoms of weakness, fever, headache, chills, sweating and muscular pain misleading doctors in diagnosis and giving correct treatment.

Long term exposure to cadmium leads to renal dysfunction with proteinuria, glucosoria, aminoaciduria, and histopathological changes in the kidney. It is also linked with the cancer of breast, lung, large intestine, prostrate and urinary bladder; hypertension, and bronchitis. The principal long-term effects of low-level exposure to cadmium are chronic obstructive pulmonary disease and emphysema and chronic renal tubular disease. There may also be effects on the cardiovascular and skeletal systems.

In 1947 in Japan, cadmium poisoning caused a mysterious disease 'itai-itai' which killed hundreds and those who survived suffered skeletal deformities. Victims were elderly women who gave birth to handicapped children who suffered deterioration in their bones. Villagers had used the contaminated water for paddy irrigation from a river near a zinc mine carrying cadmium. (UNEP and WHO Reports, 1992-2006).

5). CHROMIUM (CR)

Chromium and its oxides are widely used because of their high conductivity and anti-corrosion properties. Hexavalent (VI) chromate compounds are of greater industrial importance. They are produced by smelting, roasting, and extraction process. Chrome pigments and chrome salts are produced from sodium dichromate and used in leather and tanning industries, in chromium plating industries, for mordant dying and wood preservatives, and as an anticorrosive in cooking systems, boilers, and oil drilling muds. Chromium is used as ferrochrome for the production of stainless steel.

Human Health Risks: It is carcinogenic and corrosive on tissues. It can damage kidney and cause 'renal tubule necrosis' on long term exposure. Exposure to chromium, particularly in chrome production and chrome pigment industries, is associated with cancer of the respiratory tract. Hexavalent chromium has been found to cause high blood pressure, iron deficiency in blood, liver disease, nerve and brain damage in animals. Chromium (VI) may also cause DNA damage.

6). LEAD (PB)

Lead has wide application in human developmental / industrial activities. It is used in car batteries, television, computer, in paints and dyes, toys and newspapers, lead pipe and PVC pipes with lead stabilizers. Just one color computer monitor or television can contain up to 3.6 kg of lead. Lead in environment also comes from lead smelters, soldering and panel beating. Food, beverages, household dust, soil and water constitute the major sources of lead exposure. The biggest industrial use of lead was as an 'anti-knocking' agent in the petrol and the automobiles exhaust from the vehicles using leaded petrol were the major source of lead in the air. However, after the introduction of unleaded petrol in the 1980s almost 90 % of lead in the air has been reduced. A study made in the U.S. suggest that in 1997, the emissions of lead from the automobiles were just about 1.7 % of those emitted in the 1970. (Nevers, 2000).

Human Health Risks: Lead has been recognized as a health hazard for at least 2000 years. Lead can be absorbed into human body by inhalation and ingestion. There is mounting evidence that lead is as a '*cumulative poison*' and that exposure to low levels of lead can have significant adverse affect on health particularly for infants and children. Airborne lead affects the central nervous system. Loss of appetite, constipation, diarrhea, severe abdominal pains, limb paralysis, headaches and tiredness are just some symptoms of lead poisoning. Lead functions pharmacologically by interfering with synaptic mechanisms of transmitter release. Lead has been found to impair with RBC maturation in bone marrow, impair neurological and intellectual development, adversely effect 'intelligence' and cause learning disabilities in children (even in very low levels in blood). Most studies report a 2- to 4-point IQ deficit for each µg/dL increase in blood lead within the range of 5 to 35 µg/dL. In older people lead can cause high blood pressure.

Prolonged exposure can lead to severe brain and kidney damage, mental retardation and reduced male fertility. The fetal brain may be particularly sensitive to the toxic effects of lead because of the immaturity of the blood-brain barrier in the fetus. Researches show that lead can even increase the risk of miscarriages and still-births and may cause birth defects. Lead is readily transferred to the fetus throughout gestation period. Lead is also classified as 2B carcinogen. The most common tumors found are of the respiratory and digestive systems. However, case reports of renal adenocarcinoma in workers with prolonged occupational exposures to lead have appeared. (UNEP and WHO Reports, 1992-2006).

Lead levels in Americans and Europeans are 2 to 3 times higher than those found in other societies. Adults absorb 5 to 15 % of ingested lead and usually retain less than 5 % of what is absorbed. Children are most vulnerable to lead. They absorb lead more than the adults and even up to 30 to 40 %. Over 1.7 million American children (8.9 % of children aged 1 to 5) suffer from lead poisoning. (UNEP Report, 2001). In Mexico City 41 % of babies have been found to have dangerous lead (Pb) levels in their umbilical cords and several babies were reported to be born without brain (anenocephaly). These babies had very high levels of lead in their umbilical cord (UNEP Report, 1992). Study has found that in Bangkok, high lead exposure reduce the average child's I.Q. by 4 points. Because lead is excreted very slowly, it accumulates in the human body. The total body burden of lead may be in the bones with the largest pool in the skeleton, and with a half-life of 20 years. Lead in the CNS tends to concentrate in the gray matter. (UNEP and WHO Reports, 1992-2006).

7). MERCURY (HG)

Inorganic mercury has wide application in industrial development from manufacture of mercury thermometers and fluorescent lamps, to the production of dental amalgam, fungicides to treat seeds, and germicidal soaps and creams. They are added to paint to prevent the growth of mould on paintwork, used in detonators for explosives, and industrially as a cathode in the production of chlorine and caustic soda. About 10,000 tons of mercury is mined every year for industrial uses. Nearly 22 % of the yearly world consumption of mercury is in the electronics industry. Mercuric oxide is used in the small electrical batteries which power torches, transistor radios, small cassette tape recorders and the ordinary everyday appliances.

Other environmental sources of mercury are their ore processing facilities and the mercury chlor-alkali plants. A survey has listed 3,000 uses for mercury, and every process using it, every product incorporating it, release a small amount into the environment.

Human Health Risks: Pregnant women, nursing mothers, and children are particularly sensitive to mercury poisoning. It can affect fetal development. *Each year in the U.S. some 630,000 children are born with mercury levels that are considered unsafe by WHO* (UNEP and WHO Reports, 1992-2006). Kidneys contain the greatest concentrations of inorganic mercury after exposure to inorganic salts of mercury and mercury vapor. Experimental studies indicate that inorganic mercury can induce 'autoimmune glomerulonephritis' in kidneys of all living species. Organic mercury is more dangerous and has a greater affinity for the brain. The inorganic mercury can be transformed into organic methyl mercury in the environment through bacterial action. It has potential to cause irreversible neurological damage and is implicated in brain damage.

In the food chain as the mercury pass from one trophic level to the other it gets biomagnified to several times e.g. starting from 1 ppm it can become biomagnified to 10 ppm while it reaches the last consumer, that is human beings. The famous Minamata Disaster of Japan in the 1950s in which thousands of fishermen who ate fish from the bay were killed due to 'wasting brain disease' was caused by methyl mercury poisoning aggravating due to biomagnification. Several suffered permanent brain damage and paralysis.

GI absorption of inorganic mercury is about 7 % whereas that of methyl mercury is on the order of 90 to 95 %. Metallic mercury vaporizes at ambient air temperature and most human exposure is by inhalation. Mercury vapor easily diffuses across the alveolar membrane and is lipid soluble with greater affinity for RBC and the CNS. Inhalation of mercury vapor may produce an acute, corrosive bronchitis and interstitial pneumonitis. It may affect the CNS and cause tremor. Within cells, mercury may bind to a variety of enzyme systems, including those of microsomes and mitochondria producing nonspecific cell injury or even cell death. Methyl mercury interacts with DNA and RNA and binds with sulfhydryl groups, resulting in changes of the secondary structure of DNA and RNA synthesis.

Biological half-time of methyl (organic) mercury is about 70 days whereas for the inorganic mercury is about 40 days. Excretion of mercury from human body is by way of urine and feces differing with the form of mercury. Renal excretion increases with time. About 90 % of methyl mercury is excreted in feces. (Klaassen and Watkins, 1999).

8). NICKEL (NI)

Nickel has wide industrial application but mainly in the metal industries and in the manufacture of dry-cell batteries, coins and jewellery. Environmental sources of nickel are nickel processing industries, combustion of fossil fuels and the incineration of waste and sewage sludge.

Human Health Risks: It has been known for over 40 years that occupational exposure to nickel predisposes humans to lung and nasal cancer. Increased rates of nasal, lung and respiratory tract cancer has been reported in nickel-refining industry workers. Human exposure may be from the use of nickel-cadmium batteries, ingestion of food and drinking water, inhalation of nickel dust in air, dermal contact with jewellery and coins. Studies indicate that nickel (Ni) is also highly 'teratogenic'. Metallic nickel combines with carbon monoxide to form nickel carbonyl ($Ni[CO]_4$) which is extremely toxic and give rise to immediate respiratory failure, cerebral edema and death. (UNEP and WHO Reports, 1992-2006).

Largest concentration of nickel has been found in lungs, with lesser amount in kidneys, liver, and the brain. It is transported in the blood plasma bound to serum albumin. It is excreted through the urine which is nearly complete in 4 to 5 days.

Chapter 16

EXPOSURE TO CHEMICALS: MECHANISM OF ENTRY INTO HUMAN BODY

Natural and manmade chemicals abound everywhere in human environment-air, water and soil but are more highly concentrated in the workplaces where they are produced and where they are used. The heaviest exposure to some chemicals can occur during industrial or agricultural activities. Significant exposure can also occur through contact with naturally occurring ores and the surrounding soil, from automobile exhaust emissions, from building and insulating materials and from chemically contaminated foods. There are four main routes of exposure-

1). INHALATION

Contaminated air is inhaled through the mouth and nose and then into the lungs. An average person breathe in and out about 12 times per minute. Each inhalation brings about 500 ml of air into the lungs (corresponding to 6 litres of air per minute at rest) together with any chemical/biological contaminants that this air contains. Over an 8 hours working day (in case of industrial workers), more than 2800 litres of air will pass through the lungs.

2). ABSORPTION THROUGH SKIN AND EYES

Chemicals that pass through the skin are nearly always in liquid form. In gaseous forms they can enter after dissolving in the moisture on the skin's surface. Chemicals that can dissolve easily in lipids will pass more readily through the skin. Organic and caustic (alkaline) chemicals can soften the keratin cells and pass through this layer to the dermis, then enter the bloodstream. Some highly toxic chemicals like 'parathion' and 'sarin' penetrate the skin without causing overt damage. Corrosive chemicals can also be absorbed in greater quantity owing to increased blood flow to the skin or destruction of the outer skin barrier.

Any chemical, in the form of a liquid, dust, vapor, gas, aerosol or mist can enter the eye. Small amount of chemicals can enter the eye by dissolving in the liquid surrounding the eye. The eyes are richly supplied by blood vessels, and many chemicals can penetrate the outer tissues and pass into the veins. The organic solvent toluene can pass through the outer layers of the eye and probably enter the blood. The majority of substances that become dissolved in tears are pass to the nose and eventually swallowed. Only in very rare case does absorption of chemicals through the eye cause acute systemic effects.

3). INGESTION THROUGH DIGESTIVE SYSTEM

All forms of chemicals – liquids, solids, gaseous, vapors, mists, dusts, smoke or fumes, directly or indirectly enter the digestive system. Chemicals can enter the stomach by swallowing them or eating with contaminated food and drink. *Nail-biting is another potential source of ingested chemicals.*

4). TRANSPLACENTAL TRANSFER THROUGH MOTHER'S BLOOD IN WOMB

Considerable amount of chemicals can enter the human body at very early stage of life while in mother's womb. If pregnant women is exposed in either of the above ways the chemicals can pass across the placenta of the mother to the fetus.

In the case of each exposure route, chemicals can enter the bloodstream and thereafter be distributed to any or all of the organs and tissues of the body. In this way, they can attack and harm organs that are distant from the original point of entry, as well as cause damage at the point of entry. Following pathway is postulated for the entry of hazardous chemicals into human ecosystem-

Figure 1 explains how the hazardous materials (toxic wastes and chemicals) from the environment can enter into human body i.e. through inhalation of volatile or particulate matters emitted in the air or ingestion of chemicals leached into soil, groundwater or surface water. Chemicals in the environment can enter into human body through the food chain (uptake by edible plants and animals) and also become increasingly concentrated in the tissues of animals and human beings higher up in a process called 'bio-magnification', and can do more harm.

Chemicals are 'bio-accumulated' in organisms (plants and animals) to a concentration that exceeds several times in the surrounding environment (air, soil and water). Fishes living in water contaminated by DDT can have DDT concentrations 10,00,000 times more than that found in water. Birds and humans then eating these fishes then bio-accumulate huge doses of DDT in their bodies. Human breast milk often contain more toxic materials than the dairy milk. At death, human bodies have been found to contain (bio-accumulate) enough toxic and heavy metals to be classified as hazardous wastes. Chemicals that are persistent, toxic and bio-accumulative (PTBs) are still more dangerous.

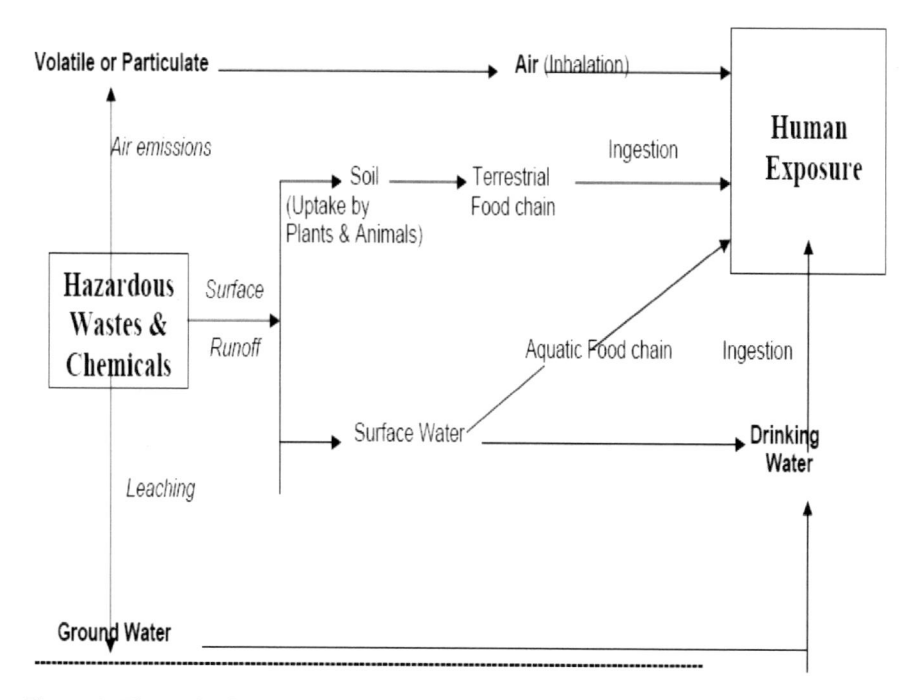

Figure 1. The Path of Entry of Hazardous Materials into the Human Ecosystem.

Chapter 17

FATE OF CHEMICALS IN HUMAN BODY

Once a chemical enters the human body, it undergoes one or more of three processes. It may be metabolized by the body's biochemical process, stored or bioaccumulated in the body and / or excreted from the body.

1). METABOLISM

Metabolism is the process by which the human body renders a 'foreign chemical' more easily excretable and less toxic with the aid of special proteins called enzymes. Enzymes often break the chemicals into simpler less toxic forms. Metabolism may occur anywhere or everywhere in human body, or in just one organ or type of tissue. For most chemicals liver is the main site of metabolism. Kidney and testes are also capable of metabolizing chemicals, the products of which may be toxic.

2). EXCRETION

Excretion is the physiological process by which unwanted chemicals are removed from the human body. Most excretion occurs through the kidneys. In the kidneys, blood carrying the foreign chemicals is filtered through a set of small twisting blood vessels called glomeruli. Some chemicals are excreted via the bile that passes from the liver into the intestine, after which the chemicals pass out of the body in the feces. Other chemicals are excreted as a gas that is exhaled from the lungs. Small amount of chemicals may also be excreted in sweat.

3). BIOACCUMULATION

Chemicals that undergo a slow metabolism or excretion are often stored / bioaccumulated in various body organs and tissues. Liver and the adipose tissues in the body fat works for storage of foreign chemicals and their metabolized derivatives if not excreted.

Half-Time or Half-Life of Chemicals in Human Body

The time taken for half of the chemicals in the human body to be metabolized or excreted is called the 'half-time' or 'half-life' of the chemical. If a chemical has a short half-time, it is metabolized or excreted quickly by the body. Some chemicals such as the heavy metal cadmium (Cd) and the pesticide DDT have a half-life of 15 to 20 years. The half-life of a chemical, however, varies from person to person, his or her immunological status, and this will influence the sensitivity of the individual to a chemical.

Chapter 18

EPIDEMIOLOGICAL STUDIES OF CHEMICALS USED IN DEVELOPMENTAL ACTIVITIES

CHEMICALS IMPAIRING HUMAN IMMUNE SYSTEMS: DESTROYING PEOPLE'S DEFENSE MECHANISM AND DISARMING THEM AGAINST INFECTIOUS DISEASES AND CANCER

The human immune system is the body's self-defense mechanism against foreign agents, but it may also play a role in containing malignant cells and thereby resisting tumor formation and cancer. Human health depends largely on proper functioning of the immune system, the foundations for which is established early in development from conception to birth in mother's womb and then from birth to year 1 and up to 18 years when immune memory is established. Maternal antibodies against several microbial diseases are transferred to the baby from week 33 to term (36th week).

Certain chemicals have been found to adversely affect the immune system impairing the immune response. It appears that immune system impairment is a potentially long-term result of altered developmental processes due to chemical exposure, the effects of which may not manifest or be recognized until later in life, long after exposure. This is the worst affect that a chemical can have on humans disarming them against the opportunistic harmful pathogens in the environment causing infectious diseases and also making them vulnerable to chemicals and radiations causing cancer.

In fact a wide range of industrial chemicals are now being identified which affect the human immune system. Better known examples are the organochlorines- the polychlorinated biphenyls (PCBs) and the dioxins. Altered T cell function has been associated with exposure to PCBs pre- and post-natal in both acute and low-level exposures. Evidence of immune suppression also exists by changes in antibody levels or immune cell numbers after exposure to chemicals.

Studies indicate that mother and children exposed to high doses of PCBs either through accidental or occupational exposures had greater incidence of respiratory and sinus infections, gastrointestinal and dermatological problems. There was increased incidence of recurrent middle-ear infections and chicken pox. In another study made on Inuit children in Quebec, Canada, those exposed to organochlorines such as the DDE (derivative of DDT) and hexachlorobenzene (HCB) exhibited higher rates of infectious diseases, particularly meningitis and broncho-pulmonary and middle-ear infections.

CHEMICALS CAUSING CANCER

Evidence accumulates of links between the proliferation of chemicals in the environment and their carcinogenic effects on human beings. It is no coincidence that the rapid rise in the incidence of cancer has occurred along with the tremendous advances in the use of chemical and nuclear technologies over the past 50 years.

WOMEN, CHEMICALS AND CANCER

A number of studies suggest that breast cancer could be related to exposure to chemical compounds like organochlorines used in the pesticides. The most common finding is an association between breast cancer risk and tissue levels of DDT and its derivative, DDE. Study in U.S. showed that levels of these substances were 50 to 60 % higher in the breast tissues of women who has cancer than in those free of it.

The millions of tones of pesticides, herbicides and other toxins- and radiation- released by industries pose increasingly complex risk to women as consumers and at the workplace. A New York State department of Health study found that women living near chemical industries ran a higher risk of breast cancer.

CHILDREN, CHEMICALS AND CANCER

Children in the modern urban society are becoming more susceptible to cancers. This is attributed to their immature immune system that takes time to mature. Children can get exposed to environmental chemicals at very early stage in their developmental history through mother's placenta. They can also get direct exposure *in-utero*. Pre-natal exposure to pesticides, nitroso compounds in cured meats and solvents have all been linked to higher risk of cancer in children.

Paternal occupational exposure to organic solvents like 'benzene' have been found to cause cancer in offspring. Early childhood exposure to pesticides used in the home appears to increase cancer risk in those children. Early exposure to asbestos increase the chances that those children will develop specific cancers as adults. It is speculated that *in-utero* exposure to persistent organic pollutants (POPs) may be responsible for increase in certain reproductive cancers among young adults. (WHO Reports, 1991-98).

CHEMICALS CAN DISRUPT HORMONE FUNCTION IN HUMAN BODY : THE LIFE'S MESSAGE

Virtually all biological development is under the control of various 'chemical messaging systems' that convey instructions from the 'genes' to their targets (cells and tissues), thereby directing development. Hormones secreted by the 'endocrine glands' play the role of 'chemical messengers'. Science has now discovered that a wide array of chemicals specially the 'persistent organic pollutants' (POPs) can disrupt this genetically based messages without damaging the genes themselves. This involves disruption of 'hormonal signaling (messaging)' through disruption of endocrine function.

Endocrine disrupting chemicals are of special health concern as they can mimic hormone produced in body. Endocrine- disrupting chemicals interfere with the activity of hormones within the human body. So far 45 chemicals have been identified which are potential 'endocrine disrupters' in humans. There is inadvertent but pervasive inclusion of 'hormonally active compounds' in several of our consumer products today- such as in many cosmetics and in plastics. Scientists are measuring the endocrine-disruption impacts of chemicals like arsenic, dioxin and bisphenol A (a basic component of polycarbonate plastic) in the low-parts-per billion. One or more message disrupting chemical contaminants have been discovered for every gland system in human body – 'thyroid system' crucial for brain development; the 'retinoid system' involved in very basic control of development; and the 'glucocorticoids' important for metabolism and tumour suppression, among other things.

Some synthetic chemicals mimic natural body hormones and send false messages. Other synthetic chemicals 'block the messages' by disruption and prevent true ones from getting through. The bottom line is that any chemical which interferes directly or indirectly with hormone function can scramble vital messages, derail development and undermine health. Hormone-disrupting chemicals may be hazardous at extremely low doses and pose an even greater danger to the developing foetus in the womb. Because hormones are crucial in early development, endocrine disrupting chemicals may alter development of child's body and brain, and undermine the ability to learn, to fight off diseases and to reproduce.

Industrial chemicals may disrupt normal function of body's 'hormonal system' causing reproductive and developmental abnormalities, neurological and immunological problems and cancer. New discoveries reveal that some 'synthetic chemicals' may mimic hormones and tend to upset normal reproductive and development processes in human beings, causing a dramatic rise in hormone-related cancers, endometriosis and other disorders. Increase in testicular, prostrate, and breast cancers have been blamed on endocrine disrupting chemicals. Such 'synthetic hormones' are now being frequently used by the meat producing industries. They pose great risk to consumers.

CHEMICALS CAUSING MALE STERILITY IN HUMANS

Certain hazardous chemicals can cause low sperm counts, structural sperm abnormalities, testicular cancer and birth defects. Dibromochloropropane (DBCP) is one such chemical. It temporarily sterilized all the men who handled it during manufacturing. Ten antibiotics including penicillin and tetracycline have been found to suppress sperm formation temporarily.

CHEMICALS AFFECTING BLOOD AND BONE MARROW

Some chemicals, such as arsine (from arsenic) damage the red blood cells (RBC), causing haemolytic anemia. Other chemicals such as benzene can damage the bone marrow and other organs where the blood cells are formed or can cause cancer of the blood-forming cells i.e. leukemia. (WHO Report, 1996).

CHEMICALS AFFECTING HEART AND BLOOD VESSELS

Some solvents such as trichloroethylene can cause fatal disturbances of the heart rhythm. Other chemicals such as carbon disulfide, may cause an increase in blood vessel disease, which may lead to heart attack. (WHO Report, 1996).

Chapter 19

PHARMACEUTICAL DRUGS AND CHEMICALS: MIXED BLESSING FOR MANKIND

Several pharmaceutical drugs contain chemicals that is proving to be a 'mixed blessing' for mankind. Some of them are life-saving drugs. Ciclosporin, an immunosupressant drug, is widely used in the prevention and treatment of graft-versus-host reactions in bone-marrow transplantation and to prevent the rejection of kidney, heart and liver transplants. It is often administered to the transplant recipients for several months and years. Ciclosporin has been linked to a remarkably high occurrence of lymphomas in the gastrointestinal tract as well as to skin cancer.

Other drugs azacitidine, chloramphenicol and chlorozotocin have also been found to be carcinogenic to humans. Dantron, widely used as a laxative, has also been found to be carcinogenic. However, the popular paracetamol used widely as an analgesic and antypyretic drug is safe for human use. Long-term experiment with paracetamol have shown reductions in tumour incidence at some sites in some animals.

SPREAD OF RADIOACTIVITY IN HUMAN ENVIRONMENT: A POTENTIAL RISK FOR HUMAN CIVILIZATION

Ever since the development of Nuclear Fission Technology for generation of 'nuclear power' several radioactive materials have been unleashed into environment. Hazardous radioactive materials resulting from nuclear fuel mining, nuclear fuel processing and operation of nuclear power plants have potential to enter into our environment and human ecosystem. Radioactive elements emitted in the atmosphere routinely from the nuclear power plants and from the nuclear explosions, such as tritium (H^3), strontium-90 (Sr^{90}), iodine-131 (I^{131}), cesium-137 (Cs^{137}) and radon (Ra^{222}) for the most part, binds to aerosols and are carried down to Earth by rainwater. Potential radiation risk to human health and environment even from the normal functioning of nuclear reactors is very high.

Every radioactive substance has a *'half-life'* (the length of time it takes for half of its radioactivity to decay and die away). Some radioactive materials with short half-lives becomes safe relatively quickly. For example iodine –131 has an 8-day half-life. After 50 days, its activity drops by over 90 %. Half-life of cesium –137 is 33 years. About 20-30 millicuries of Cs^{137} accidently leaked out from the Tarapur Atomic Power Plant in Maharashtra (India). Half-life of radium-226 is 1600 years. All nuclear reactors produce radioactive wastes with half-life of 24,000 years. The *half-life* of uranium–238 is 4.5 million years and its fission product plutonium-239 also has very long *half-life*. Even after 50,000 years they will have lost only three-quarters of its radioactivity and still be lethal.

Radioactive substances (can be termed as 'radioactive wastes' in case of the nuclear power plants) are produced regardless of whether nuclear fission is controlled (such as for energy generation in reactors), or occurs explosively, as in the atom bomb. The resulting fission products, isotopes of approximately 30 elements, have mass numbers in the range of 72 to 162, are for the most parts solids, and emit beta particles, together with electromagnetic reaction (gamma rays) which are exceedingly penetrating. The chemical separation of fission products and their conversion to nuclear fuel are the most important sources of radioactive materials in the environment. The radioactive wastes can be in all the three forms – solid, liquid and gaseous and two categories of radioactive wastes are mostly encountered- the Low Level Radioactivity Waste (LLRW) and the High Level Radioactive Wastes (HLRW). Eight tons of liquid radioactive waste result per year from the typical average-size, nuclear reactor. Another potential source of radioisotopes in the reactor water is the fission products formed within the fuel elements.

The 'nuclear reactors' mostly generate HLRW in the form of plutonium-239 (Pu^{239}). Other two most significant fission products are strontium-90 (Sr^{90}) and cesium-137 (Cs^{137}) with half-life of 19.9 and 33 years respectively. They are routinely emitted from the reactors and continue to release radiation energy over long periods of time (several generations of the human race).

Table 6. Half-life of Some Radioisotopes Commonly Used in Developmental Activities

Mn^{56}	2.6 hours
Cu^{64}	12.8 hours
I^{131}	8 days
Fe^{59}	45 days
Zr^{95}	65 days
Cs^{137}	33 years
Sr^{90}	19.9 years
Ra^{226}	1600 years
U^{238}	4.5 million years

Source: From Publications of UNEP and WHO (1992 – 2006).

Even dismantling (decommissioning) of retiring nuclear reactors produce enormous amount of radioactive wastes and contaminate vast land area. Dismantling produces about 150 million cubic feet of Low Level Radioactive Waste (LLRW) which is several times more what is produced each year by all the world's operating reactors. There are about 435 nuclear reactors in world mostly in France and Japan. Ambitious nuclear power program was launched in the US, UK, Canada, former USSR, France, Germany, Japan, Belgium, Sweden and also in India and Argentina. France and Japan today are the biggest producers of nuclear power meeting 70 % of their needs. The old reactors have produced mountainous nuclear wastes including over 1400 highly radioactive spent fuel rods.

Uranium mining and processing (enrichment) produces huge amount of solid and liquid radioactive wastes which is highly hazardous. The extraction of uranium from the earth crust leaves vast quantities of wastes as 'tailings' which contain up to 80 % of the original radioactivity of the extracted ore. After mining, uranium is further enriched to produce 'nuclear fuel'. Depleted uranium hexafluoride (DU) is a radioactive waste by-product of enrichment. Processing of uranium ores produces considerable volumes of alpha emitters, mainly radium (Ra) and gives rise to a toxic gas 'radon'. For every 1000 tones of processed uranium fuel, 100,000 tones of mined wastes as tailings and 3,500,000 liters of liquid waste is produced. They migrate into the environment through air, soil and water.

High-level radioactive wastes (HLRW) are generated during fuel reprocessing and it is a highly risky and hazardous process. Solid nuclear fuels, which have been stored for about 3 to 4 months, are dissolved in nitric acid (HNO_3), and fed to the extraction column, where uranium and plutonium are extracted with a particular chosen solvent. To remove trace of fission products, the extracted uranium and plutonium are scrubbed by an aqueous salt solution added at the top of the extraction column. The waste solutions, which contain more than 99.9 % of the total fission products and inert in the feed and the scrub, leave at the bottom of the column. Normally the two most significant fission products are strontium-90 (Sr^{90}) and cesium-137 (Cs^{137}). The 'Friends of the Earth-Brisbane' reports that Australia produces around 14,000 tones of uranium 'tailings' each day which contains up to 80 % of the original radioactivity of the extracted ore.

NUCLEAR ACCIDENTS AND GLOBAL SPREAD OF RADIOACTIVITY

Several nuclear reactor accidents in the past have left an ugly legacy of radioactive hazardous wastes on earth. Minor nuclear accidents are of common occurrence. There were 300 minor nuclear accidents in Germany in 1988. Several nuclear accidents in the past have killed many workers, while crippled others who survived, for their whole life.

1. In 1957, the overheating of nuclear reactor at Windscale, Cumberland, UK, spewed radioactive Iodine 131 (I^{131}) over an area of 800 square km of land contaminating it. As a consequence, all milk produced in the area was declared 'unsafe', barred from shipment, and dumped. At least 20 workers died of cancer.
2. In 1969, in Colorado, U.S., spontaneous ignition in a nuclear waste pile caused a massive release of plutonium-239 dust in the air.
3. In 1976, again the nuclear reactor at Windscale, U.K. accidently leaked 2 million litres of radioactive wastewater.
4. In 1979, a serious accident occurred at the 880 MWe Three Mile Island Nuclear Power Plant in the U.S. Large amount of radioactive fission products consisting mainly of noble gases 'Xenon-131', 'Xenon-135' and traces of Iodine-131 were released which continued for months.
5. In 1986, the Chernobyl Nuclear Disaster in Ukraine (erstwhile USSR) was worst in developmental history of NPP. It released about 30 radionucleides with a total radioactivity of about 2900 PBq. Clouds of radioactive debris hurled in the sky for several days and the dust were

carried away by the wind to other nations. In two weeks the nuclear fallout covered 20 countries of the northern hemisphere. 135 tones of uranium, plutonium and other radioactive elements lie buried and entombed at the site of disaster which will remain alive for at least 100,000 years. Deadly radioactive elements fell to Earth in nuclear rain and passed into the food chain of grazing animals. Milk and butter were found to be contaminated and large number of cattle had to be slaughtered and incinerated.

After the explosion and to prevent the further spread of radioactivity into the environment, the remains of the old reactor at Chernobyl was covered by a concrete structure which is 248 feet high and named as 'SARCOPHAGUS'. Government claimed that the structure would last like the Egyptian 'pyramids'. But as it was all done through a remote-control technology the shelter is not completely air-tight and large numbers of cracks and openings have appeared. There is high apprehension that the structure may collapse by a mild tremor or heavy rainfall and landslides emitting a lethal cloud of up to 50 tons of radioactive dust repeating the events of 1986.

6. In 1986, in Oklahoma, U.S., a tank at the uranium reprocessing plant ruptured releasing about 14,000 pounds of slightly radioactive 'uranium hexafluoride gas' which breaks down into toxic 'hydrogen fluoride' and low level radioactive 'uranyl fluoride' particles. The gas rise to the land surface from any source of uranium and can travel horizontally for several kilometers before penetrating in the cracks of house foundations. (WHO and UNEP Reports, 2002).

7. In 1999, a nuclear accident occurred in Uranium Processing Plant in Tokaimura, Japan exposing its workers to high doses of radiation. The radiation kept leaking for more than 20 hours before it was brought under control. It left 320,000 people sheltering in their homes due to threat of spread of radiation in the area.

POTENTIAL HEALTH IMPACTS OF RADIOACTIVE MATERIALS IN THE ENVIRONMENT

Since the radiations are capable of altering the atoms of matter through which they pass and have a 'cumulative effect' in nature, they may cause irreparable damage to living tissues. The International Commission on

Radiation Protection (1979) has established limits of maximum permissible concentration (MPC) for the United Nations. MPC for general population is 5 roentgens over a 30-year period.(One roentgens (r) is the quantity of radiation which causes one gram of living tissue to absorb 97 ergs of energy). The commission recommends that no member of the society receive a dose of radiation to exceed a total of 500 millirems per year over a lifetime.

IMPENDING DEATH FOR RADIATION VICTIMS

People working in uranium mining industry and in the nuclear power plants, and those living in its vicinity is always at risk. Potential risk to human health and environment even from the normal functioning of nuclear reactors is very high. There is routine emission of radioactive isotopes (like Strontium-90, Cesium-137 and Iodine-131 which has cumulative effect on human body) when the reactor is in operation and major concern has grown about the health and safety of workers. The ingestion of radioactive products from the use of radioactive water in industries can have a somatic effect on human beings, causing malignant tumours, or chromosomal and gene (heredity materials) mutations that might affect the future generations.

After the Chernobyl Nuclear disaster in 1986 large areas of Ukraine and Bylorussia in the former USSR were declared unfit for human habitation and people were foredoomed to slow agonizing deaths in the days, weeks, months and years to come. Although only 31 workers perished at the site of disaster tens of thousands of people in the countries of former USSR and the rest of Europe were destined to die impending death. Among the workers who survived Chernobyl disaster, there is 50 % higher incidence of thyroid cancer, leukemia, stillbirths, malignancy, miscarriages and genetical deformations. Among the population living around Chernobyl where 250,000 people have already died, there is 50 % higher incidence of thyroid cancer, leukemia, still births, malignancy, miscarriages and genetical deformations. Many children are terminally ill even today, and by 1993, 12,000 of them had died. (UNEP Report 2002).

After the nuclear accident in Uranium Processing Plant in Japan (1999), three workers who were exposed to radiation doses 20,000 times the normal levels died within weeks, while 49 were exposed to high levels of radiation. They faced *'lingering illness'* and impending death in the coming years. The 3 workers closest to the nuclear accident started vomiting and diarrhaea immediately. Shortly after they had headache, fatigue, dizziness, and internal bleeding. This was followed by fever and low B.P. Within a week immune system collapsed and internal organs shut down. Emergency workers in the plant who carried the injured were also contaminated.

Main threat of radiation arise by contamination of water used for human consumption or recreation, fish food or the edible crop plants irrigated by contaminated water. Some radioisotopes like strontium-90 (Sr^{90}) and cesium-137 (Cs^{137}) are routinely emitted from the nuclear reactors. Once in the environment they behave like the nutrients calcium (Ca) and potassium (K) respectively and pass into human food chain from soil → grass → cattle milk → human body. Sr^{90} has affinity for bone marrow. Significant amount of Sr^{90} was found in the bone marrow of infants suffering from leukemia in Maharashtra state of India. Cs^{137} has affinity for soft tissues where it can cause us to age more quickly. Recently iodine-129 (I^{129}) with half-life of 1.6 years has also received attention. It can replace normal iodine in water and get lodged in the thyroid (iodine is important for thyroid function) and may induce thyroid cancer. Plutonium-239 (Pu^{239}) mimics iron and finds its way into bone, blood and reproductive organs.

In the Oklahoma accident (1986) where about 14,000 pounds of radioactive 'uranium hexafluoride gas' penetrated in the cracks of American homes, it is feared that as many as 30,000 lung cancer deaths every year may occur due to contamination by the radon gas. After the Gulf War of 1991 where DU tipped bullets were used there is significant increase in congenital deformities in babies born in Kuwait and southern Iraq. Higher incidences of cancers have been reported from the region of Nevada desert in the Pacific Ocean where US conducted nuclear tests. Among the sufferers are many Pacific Islanders.

The fallout from the nuclear test conducted by UK in the Monte Bello Island off Western Australia spread to 3,200 kms to the east where the level of radioactive Iodine-131 was 100 fold. Background radiation in Melbourne and Adelaide were raised more than 100 times. Nuclear tests were conducted in the 1950s at Maralinga in South Australia. Half a century later, it is still not safe for the indigenous people to return to their homeland. All lakes, streams and water-bodies are contaminated. Significant level of radiation still persists in the environment and the ecosystem of Pokharan in the Thar Desert of Rajasthan (India) where underground nuclear test was conducted in 1975. There were lots of reports of radiation-related damages (still-births, birth of handicapped children etc.) from the population living in the area.

Table 7. Accidental Radiation Effects on Man

Dose of Radiation Absorbed	Health Effects
1. Less than 25 rads	No observable effects
2. About 25 rads	Threshold Level
3. About 50 rads	Slight (temporary) changes in blood
4. About 100 rads	Nausea, vomiting and fatigueness
5. Between 200 – 250 rads	Fatality possible, though recovery is likely
6. About 500 rads	Perhaps half the victim would die
7. About 1000 rads	All the victims would die; death would occur within days
8. About 10,000 rads	Death within hours
9. About 1,00,000 rads	Death within minutes

Source: UNEP (Rad is defined as that quantity of radiation which leads to absorption of 100 ergs / gram of the absorbing material).

Chapter 20

SOME IMPORTANT GLOBAL ACTIONS FOR REMOVAL OF TOXIC CHEMICALS FROM HUMAN ENVIRONMENT: THE GREEN CHEMISTRY MOVEMENT

A Green Chemistry Movement (GCM) initiated by Paul Anastas of U.S. is going across the world since 1990 after the passing of the Pollution Prevention Act by EPA. This has several objectives to reduce or even eliminate the use of toxic chemicals in production process. Principles of green chemistry includes

1) Avoid or minimize using solvents, separation agents, or other auxiliary chemicals and if necessary use safer solvents and water as solvent, and safe reaction conditions;
2) Design safer chemicals and chemical products that have 'little or no toxicity' and which can break down into innocuous substances after use and do not accumulate in the environment;
3) Design less hazardous chemical synthesis by maximizing atom economy so that the final product contain the maximum proportion of the starting materials without wasting much atoms into chemical wastes;
4) Use 'renewable feed-stocks' for chemical production such as 'corn' and 'soybeans';
5) Use chemical catalysts to promote chemical reactions instead of stoichiometric reagents because catalysts are required in very small amounts and can carry out a single reaction many times as compared to the stoichiometric reagents which are used in excess and work only once;

6) Increase energy efficiency by catalyzing chemical reactions at ambient temperature and pressure whenever possible;
7) Avoid chemical derivatives by preventing use of blocking groups because derivatives use additional reagents and generate more chemical wastes. (Anastas and Warner, 1998).

An environmentally friendly 'microwave mediated' organic chemical reaction has been developed which is more rapid and safe with high yields and use very little synthetic solvent and even use water as a solvent. There is very little or no chemical waste resulting. A new technology called 'catalytic dehydrogenation of diethanolamine' has been developed in herbicide producing industries that avoids the use of toxic chemicals cyanide and formaldehyde. It is also safer to operate, produces high yield and has fewer process steps. (Anastas and Williamson, 1998). Supercritical carbon dioxide is now substituted for perchloroethylene as a solvent in professional dry cleaning. Water has replaced the 'petroleum distillates' in paint industries. Manufacture of 'ibuprofen' no longer creates 'toxic cyanide' and 'formaldehyde' as hazardous chemical wastes.

SOME LEGISLATIVE ACTIONS

International Register of Potentially Toxic Chemicals (IRPTC) (1976)

The IRPTC was established by UNEP in 1976 to collect and disseminate global information on hazardous chemicals, including national and international laws and regulations controlling their use. IRPTC operates through network of national and international organizations, industries, external contractors, and national correspondents. IRPTC computerized data file contains profiles of over 800 potentially toxic chemicals, and 8000 potentially toxic substances. It also provides information on hazardous waste management and safe disposal to industries.

Toxic Substances Control Act (1976)

This was enacted as early as in 1976 to stop production of toxic synthetic chemicals and search for benign and sustainable alternatives for uses in production process.

International Program on Chemical Safety (IPCS) (1980)

IPCS was set up in 1980 by WHO, UNEP and ILO to assess the risk that specific chemicals pose to human health and the environment. IPCS publishes its evaluation in four forms-

1. Environmental Health Criteria for Scientific Experts;
2. Health and Safety Guides for General Public;
3. Chemical Safety Cards for Ready Reference at Workplace for the Workers;
4. Poison Information Monographs for Medical Use.

EU Legislation on 'Registration, Evaluation and Authorization of Chemicals' (REACH)

European Union is moving towards a new legislation called 'Registration, Evaluation and Authorization of Chemicals' (REACH). The use of hazardous chemicals that disrupt endocrine function, have risk of causing cancers, mutations (genetic deformations) or problem with reproduction, or those that bio-accumulate in human body much faster than others- will require a specific permit for production or may be prohibited altogether. This will send a clear message to all industry- look for, and develop, safer chemicals in developmental activities. (UNEP, 2006).

The International Chemical Agenda at the Rio Earth Summit (1992)

Considering the 'close linkage' between the chemicals, environmental toxicity and human health an integrated global response to eliminate toxic chemicals from the human environment began during the United Nations Conference on Environment and Development (UNCED) in Brazil in 1992, called the 'Rio Earth Summit'. The Chapter 19 of 'Agenda 21' (Environmental Agenda for the 21st Century) of the Rio Declaration is the framework upon which the International Chemicals Agenda (ICA) is built. It called for a globally harmonized 'hazard classification' and 'compatible labeling system'- including material safety data sheets and easily understandable 'symbols' for toxic chemicals and wastes. Subsequently, an 'Intergovernmental Forum on

Chemical Safety' (IFCS) was created in Stockholm in April 1994. It is a mechanism for cooperation among governments for promoting 'risk assessment' and environmentally sound management of chemicals. The Inter-Organization Program for the Sound Management of Chemicals (IOMC) was established in 1995. It serve as a mechanism for coordinating efforts of international and intergovernmental organizations. Together they involve the United Nation Environment Program (UNEP), the World Health Organization (WHO), the International Labour Organization (ILO), the Food and Agriculture Organization (FAO), the United Nation Industrial Development Organization (UNIDO), the United Nation Institute for Training and Research (UNITR) and the Organization for Economic Cooperation and Development (OECD), which works like the mortar to bind the elements together. They provide the forum where government and non-government stakeholders could address issues on 'toxic chemicals'. The IFCS recommended to evaluate 200 chemicals by 1997, and another 300 by year 2000 which has now been completed. It adopted the 'Priorities for Action Beyond 2000' called the 'Bahia Declaration' to further advance the works of Agenda 21.

UNEP has published *'Legislating Chemicals : An Overview and Code of Ethics on the International Trade in Chemicals'*, which provides guidance to governments and industry on legislation to enhance management, and helps countries to develop national systems to strengthen the control of 'unregulated chemicals'. The UNITR, in cooperation with the UNEP, FAO, UNIDO, OECD and the WHO, is developing and implementing several 'training and capacity-building' programs on chemical management. A joint declaration of UNEP and FAO imposing restriction on export of some toxic chemicals such as 'polybrominated biphenyls', 'polychlorinated biphenyls', 'polychlorinated terphenyls', 'tris (2 and 3 dibromopropyl) phosphate', and 'crocidolite'. Several countries have imposed total ban on the use of these chemicals and products derived from them.

The Rotterdam Convention on the Prior Informed Consent (PIC) Procedure for Certain Hazardous Chemicals and Pesticides in International Trade (1998)

Adopted in 1998 and jointly administered by United Nation Environment Program (UNEP) and Food and Agriculture Organization (FAO), it provides a first line of defense in controlling the risks from hazardous pesticides and chemicals. It addresses the international trade in chemicals, and provides a mechanism whereby developing countries can keep dangerous substances from being exported into their territory. It strengthen the power of importing countries to decide which chemicals they want to receive and to exclude those they cannot manage safely. It also ensures that exporting countries will respect those decisions.

The Rotterdam Convention also promotes international cooperation by requiring the governments of the developed nations to provide technical assistance to developing countries. The Convention was adopted in Rotterdam in September 1998 and was initially signed by 61 countries. At its adoption, the Convention covered 27 chemicals. Hundreds more are likely to be added. Parties to the Convention agreed not to export 27 harmful pesticides and industrial chemicals that have been banned or restricted (especially those causing problems because of the way they are used in the developing countries) unless the importing countries agrees to accept them.

Representatives of 17 developing countries of Asia met in Bangkok in June 1999 to discuss limits on trade in certain hazardous chemicals and pesticides. The purpose of the meeting was to identify steps they must take to implement the Rotterdam Convention.

By the end of 2001, the Convention had been signed by 73 states and the European Union. Till date more than 17 governments have ratified and others are in process. The Convention incorporates the Prior Informed Consent procedures that 154 countries have undertaken voluntarily since 1989. The Convention came into force on 24th February 2004.

Substances subject to the Convention's requirement include the following pesticides-

1. 2,4,5-T
2. Aldrin
3. Captafol
4. Chlordane
5. Chloridmeform
6. Chlorobenzilate
7. DDT
8. Dieldrin
9. Dinoseb
10. 1,2-dibromoethane (EDB)
11. Fluoroacetamide
12. Hexachlorocyclohexane (HCH)
13. Heptachlor

14. Hexachlorobenzine
15. Lindane, and
16. Mercury compounds

It also include certain formulations of

17. Monocrotophos,
18. Methamidophos,
19. Phosphamidon,
20. Methyl-parathion, and
21. Parathion.

Industrial chemicals included in the Convention are –

1. Crocidolite asbestos
2. Polybrominated Biphenyls (PBBs)
3. Polychlorinated Biphenyls (PCBs)
4. Polychlorinated Terphenyls (PCTs), and
5. Tris (2,3 dibromoprophy) phosphate.

The first Conference of the Parties to the Rotterdam Convention met in Geneva in 2004 adding 14 additional chemicals to the Convention, and decided to make the legally binding PIC procedure operational.

The Stockholm Convention on Persistent Organic Pollutants (POP's) (2001) : Banning the Dirty Dozen

In February 1997, UNEP's Governing Council meeting concluded that there was sufficient scientific evidence to ban the production of some of the very dangerous industrial chemicals identified as the 'Persistent Organic Pollutants' (POP's). They persisted in the environment for exceptionally long period of times, and traveled long distances, even up to the Arctic poles of Earth thousands of kilometers from any major source of POP's, and were detrimental to human health even in very low concentrations. POP's released in one part of world are transported through the atmosphere by a repeated process of evaporation and deposit, re-evaporation and re-deposit. They circulate globally through a process called 'grasshopper effect'.

UNEP adopted Decision 19 / 13 C which urged governments to take immediate action on the Persistent Organic Pollutants (POPs). The World Health Assembly of the World Health Organization (WHO) also endorsed the action. Accordingly a series of 8 regional and sub-regional workshops were held in 1997-98. 138 countries attended the workshops and a treaty was negotiated under the United Nation Environment Program (UNEPs) auspices in Montreal, Canada in June 1998. The intergovernmental negotiations were concluded in Johannesburg, South Africa on 10 December, 2000.

The treaty was finally adopted on 22nd May 2001 in Stockholm, Sweden, and became a Convention. A decision to ban the production of 12 'Persistent Organic Pollutants' (POPs) was taken. It currently covers 12 chemicals (causing death and birth defects) viz. aldrin, dieldrin, endrin, heptachlor, chlordane, hexachlorobenzene, mirex, polychlorinated dioxins and furans, DDT, PCBs, and the toxaphene, but has a process to add new POPs. The convention provides the means to prohibit their production and use, and to reduce and – where feasible- ultimately eliminate their release to the environment. It also contains a financial mechanism. Although most are subject to immediate ban, DDT and PCBs have been given some exemption. DDT is still needed by some developing countries to fight malaria.

PCB-containing materials will have to be inventoried, removed from services by 2025 and properly stored and then disposed off in an environmentally sound manner by 2028. PCBs have been used since the 1930s in a wide range of applications, including in electrical equipment and transmission systems such as in transformers and capacitors. Although PCB's are no longer produced, hundreds of thousands of tones are still used in such equipments.

International Persistent Organic Pollutants (POPs) Elimination Network

An International POPs Elimination Network (IPEN) has been formed under the auspices of UNEP which now includes over 400 environmental groups (NGOs) committed to taking action on banning the production of chemicals detrimental to human health and the environment.

EU Commission's White Paper on Strategy for a Future Chemical Policy (2001)

The European Union Commission (EU) is currently the most progressive international proponent of environmentally sound management of toxic chemicals. The vast majority of the 100,000 chemicals potentially produced and used in the European Union (EU) were placed on the market before new EU assessment procedures came into force in 1981. EU's White Paper - *'Strategy for a Future Chemicals Policy'* (2001) represent a responsible step towards ensuring that the chemical industries provides at least some data on all the chemicals produced in quantities over 1 ton. It will include-

(1) A coherent system for existing and new chemicals to ensure that sufficient information is available for all chemicals;
(2) A responsibility to generate data about chemicals, on the part of industries those who produce and market them;
(3) Registration of all chemicals above a certain production tonnage per year;
(4) An authorization scheme for the use of chemicals of very high concern for which no alternative yet exist.
(5) Transparency of information about chemicals and free public access to it.

The first priority is to close the knowledge gap on the risks associated with the estimated 30,000 existing chemicals of which over 1 ton is produced or manufactured per manufacturer per year. The threshold does not mean that hazardous substances produced or imported in quantities under 1 ton will 'escape' control. Industry will have to take responsibility and will be required to assess the toxicity and health hazards of each substance and provide information to users in the form of 'labels' and 'safety data sheet'. Unfortunately the EU's proposal only addresses the un- POPs-like substances and those that are carcinogenic (inducing cancer), mutagenic (inducing genetical changes) or reprotoxic (interfering with reproductive processes). WWF has advocated to include a range of chemicals in the EU's policy.

Many local authorities in Europe have PVC-free policies for municipal buildings, pipes, wall-papers, flooring, windows and packaging. Recent concerns about the use of softeners in PVC plastic toys which has potential to leach out into mouth of children's have led to further restrictions on PVC use in Europe.

EU Directives on 'Restriction on Hazardous Substances' (ROHS),

In January 2003, the EU Parliament enacted Directives on *'Restriction on Hazardous Substances'* (ROHS), calls for phasing out of heavy metals mercury (Hg), cadmium (Cd), lead (Pb) hexavalent chromium (Cr VI) and the two classes of brominated flame-retardants e.g. (Polybrominated Biphenyls and Polybrominated Diphenyl Ethers) in all electronics and electrical items by July 2006 (with a number of exemptions) which has now come into effect. The Directive encourages research into alternative materials which are good substitutes and also environmentally benign.The 15 Member States of the EU in general welcomed the Directive which entered into force in 2006. Netherlands, Denmark, Sweden, Austria, Belgium, Italy, Finland and Germany are already ahead with legislation.

Swedish Policy on Phasing out of Hazardous Chemicals from New Products

Swedish Policy on hazardous chemicals is several steps ahead of others. It requires phasing out of hazardous heavy metals like lead, mercury and cadmium, and all the potential carcinogenic, reprotoxic and mutagenic chemicals from its new consumer products wherever possible.

In May 1998, Sweden's National Chemicals Inspectorate called for a ban on Polybrominated Biphenyls (PBBs) and Polybrominated Diphenylethers (PBDEs) while urging their government to work for a European wide ban and for controls on the international trade in these chemicals. As a consequence, PBBs will no longer be used in Sweden commercially. German chemical industry has already stopped production of these chemicals in 1986.

The Technological Actions

Development of Safer Chemicals in Chemical Industry
Perhaps the greatest challenge to the chemical industries of world today is to design and develop environmentally safer chemicals that :

1. Do not 'bio-accumulate' in the living systems;
2. Do not have 'latent effects' or 'cumulative' effect' of exposure after elapse of several years;
3. Can 'biodegrade' rapidly and completely in the environment;
4. Do not build resistance in organisms.

The plant-based biochemicals qualify for all the above criteria and are emerging as environmentally safer alternative to the mineral based natural and manmade synthetic chemicals. (Sinha and Herat, 2005).

Removal of Toxic Chemicals from Production Process

People do not attach significance to the material but the services they get. Meeting peoples need of materials which is 'non-chemical' or a 'non-toxic' chemical or 'less-toxic' material or 'most appropriate materials available in the environment' whose procurement from the earth's crust entails minimum environmental damage, is the key to protection of environment and human health.

Safer alternative to dangerous chemicals like methyl isocyanate, oleum, ethylene, butadiene, propylene, benzene and vinyl chloride used in various developmental activities are being developed. Use of chemicals sometimes becomes inevitable in industrial processing to get a valued product. There is need to search and develop 'alternatives' and 'substitutes' for such chemicals / materials that are 'lesser evil' or none at all. There are several chemicals and substances in the environment which are environmentally benign and can replace the existing toxic substances used in the industries as their 'substitutes' and 'alternatives'. (Sinha and Herat, 2005 b).

Eliminating pigments containing heavy metals from ink and paint formulations; replacing chlorinated solvents with non-chlorinated solvents in cleaning products; replacing phenolic biocides with less toxic compounds in metal-working fluids; and developing new paint, ink and adhesive formulations based on water rather than organic solvents are good examples of material change. (Anastas and Williamson, 1998).

Some Examples of Use of Benign Chemicals in Industrial Production Process

1) In electronics industry toxic organic solvents used to de-grease metal surfaces and circuit boards have been replaced by 'citrus-based'

vegetable oils or just soap and water – all biodegradable products completely eliminating toxicity in wastewater.
2) In dyeing and printing industries the dangerous chemical methyl-ethyl ketone (MEK) is now being replaced. Vegetable dyes like indigo and myrobalan which were used in traditional systems of dyeing and printing in India are now being revived.
3) In manufacturing household appliances the 'vapor degreaser' using chlorinated solvents is replaced by safer 'alkaline degreaser';
4) In manufacturing electronic components ozone can replace the toxic organic biocide for killing the algal weeds in cooling towers;
5) In manufacturing plumbing fixtures the 'hexavalent chrome-plating bath' is replaced with low-concentration 'trivalent chrome-plating bath';
6) In manufacturing ink cadmium (Cd) pigments has been removed from the ink.

Table 8. Environmentally Benign Substitutes of Some Toxic Materials / Chemicals Used in Industries

Toxic Substances	Benign Substitutes
1. Arsenic	Synthetic Organic Chemicals
2. Asbestos	Glass fibers
3. Lead	Aluminum, Tin and Plastics
4. Cadmium	Tin and Zinc
5. Mercury	Lithium
6. Nickel	Plastic
7. Silicon / Copper (Ore)	Optical Glass Fibers / Conducting Plastics
8. Iron and Steel (Ore)	Carbon Fibers / Durable PVC / Ceramics

Source: World Resource Institute (WRI), Washington (1986).

Reducing or eliminating use of hazardous materials / chemicals in production process will decrease not only the generation of hazardous chemical wastes, but also the amount of gaseous hazardous materials in air emissions and liquid hazardous chemicals in wastewater effluents.

Material change may require some minor process adjustments or it may require extensive new process equipment which is paid back by way of reduced waste and pollutants.

Table 9. Examples of Change of Hazardous Materials in Industrial Production

Industry	Change of Hazardous Materials
1. Manufacturing Household Appliances	Vapor Degreaser using chlorinated solvents can be replaced by safer 'alkaline degreaser'
2. Manufacturing Electronic Components	Ozone can replace the toxic organic biocide for killing the algal weeds in cooling towers.
3. Manufacturing Plumbing Fixtures	The hexavalent chrome-plating bath can be replaced with low-concentration trivalent chrome-plating bath
4. Manufacturing Ink	Cadmium (Cd) pigments has been removed from the ink

Source: Nemerow, Nelson *'Zero Pollution Industry'* (1995).

Replacing Organic Solvents by Water-Based Solvents

Water can replace several organic industrial solvents used in the production process specially where ink, dyes, adhesives and paints are manufactured and reduce toxic wastes.

1) In printing industries, the solvent-based ink can be substituted by water-based ink. In ink manufacture the cadmium pigments can be removed from the product. In Denmark the Dansk Transfertryk company has developed a technique for printing with water-based dyes on synthetics, and improved process for printing on cotton. (UNIDO, 1994)
2) In industries manufacturing printed circuits boards water-based developing system can replace solvent-based developing systems. A circuit-board manufacturer switching to water-based flux from a solvent-based product will have to replace the solvent-vapor degreaser with a detergent parts washer. This may involve some limited capital investment.
3) One firm in the U.S. just used a household dishwasher to replace the solvent-vapor degreaser. Producers of gift-wrapping papers in U.S., simply switched from solvent to water based inks.

4) In Australia, an innovative technology has been developed to clean the greased machinery parts in water by passing 'sound waves' of particular wavelength through the objects placed in water. There is no use of toxic chemicals and no chemical waste is therefore discharged. The water is also recycled for reuse.

5) Painting industries use solvent based paints containing glycol, formaldehyde, benzene, ethers, ammonia etc. and several chemicals as cleaning agents for surface preparation. Water based paints is replacing the solvent based paints for use in homes and indoors. Australian paint company Rockcote have produced seven eco-friendly products which are water-based and free of dangerous chemicals. These are – high gloss water-borne enamel, semi gloss water-borne enamel, sealer / undercoat, low sheen, ceiling white, Torino suede effect wall coating and Portofino fine sand or adobe finish. (Rockote Australia, 2003).

Production of Chemical-Free and Biodegradable Agro-plastics and Polyesters

Plastics are most convenient materials used by the people today. Biotechnology has developed a new 'biodegradable polyester' called 'politri methylene terephthlate'. It has been made by fermentation of carbohydrates from corn, beet and potato and agricultural wastes. A single microbe has been genetically tailored which possess all the enzymes for conversion of sugar into 'glycerol' and to 'polyester'. The fibers are heat-settable and stable to moisture. There is no use of heavy metals, petroleum products or any toxic chemicals in the process and no generation of hazardous chemical waste. Bacteria can produce biodegradable plastics through fermentation but the process is five times more expensive than the production of conventional non-biodegradable plastics made from petroleum. The Central Tuber Crop Research Institute (CTCRI) in Kerala, India is has produced a biodegradable plastic by mixing low density polyethylene with starch. Bacteria would eat up the polyethylene with starch. Monsanto of U.S. has produced a biodegradable agro-plastic PHBV (poly 3-hydroxybuyrate-co-3-hydroxyvalerate) from cress and oil seed rape plants through genetic manipulation of four bacterial genes and the metabolic pathways of amino acids and fatty acids synthesis. Commercialization will take time but the door has been opened. (UNEP Reports, 1996-2004).

Production of Conducting Plastics without Use of Toxic Chemicals

A conducting plastics made from polyacetylene (PA) has heralded a new era as an environmentally benign substitutes for the conducting metals - copper and silicon. The most commonly proposed application of 'conducting plastics' is their use as electrodes in light weight and rechargeable batteries. PA batteries do not use toxic materials (Sinha and Herat, 2005).

Biochemicals Can Replace Toxic Agrochemicals for Safe Food Production

A new bio-pesticide for sugarcane called 'Bio-Cane' has been developed in Australia to replace the chemical pesticide. It is a fungal product cultured on broken rice grains. The biocane granules are particularly effective against 'greyback canegrub' disease in sugarcanes. Several plants of Solanacea family have been found to produce a new class of chemicals called 'withanolides'. They act as repellents, feeding deterrents and toxins to a wide variety of pests. Several plants have come to light that have potent 'insecticidal' and 'pesticidal' properties. 'Azadirachtin' from the Indian margosa tree (*Azadirachta indica*) isolated by the Max Planck Institute of Germany. Like chemical pesticides the herbal pesticides do not kill the 'natural enemies' of pests and also do not induce 'resistance' in organisms. Other potent herbal chemical pesticides are 'pyrethrins' isolated from *Chrysanthemum cineriifolium* and 'rotenoids' from *Tephrosia villosa, Derris elliptica* and *Parthenium argentatum*; 'napthaquinones' from *Plumbago capensis*, 'amorphin' from *Amorpha fruticosa*. Pyrethrin is a biochemical more deadly to insects than DDT, yet hardly toxic to humans. It is also non-persistent and hence environmentally safe. It can be safely used as an insecticide in the house where food is processed and stored. The plants are now grown widely in uplands in Rwanda, Kenya, Tanzania, Ecuador, Japan and Australia. (Sinha and Herat, 2005).

Cleaner Production in Some Consumer Industries : Substituting With Benign Chemicals or Eliminating Use of Toxic Chemicals

1). Paper Industry

1) Change from 'calcium hypochlorite' to 'sodium hypochlorite' in bleaching process;

2) Elemental chlorine-free bleaching and total chlorine-free bleaching. An environmentally benign oxidant 'hydrogen peroxide' is being used for bleaching. Enzymatic bleaching of pulp reduce and even eliminate the need for hydrogen peroxide;

NIRO-Separation Company of Denmark has developed a process for making paper without the use of water and chlorine. The pulp is dissolved in air and transported through the process by means of air. It is made into paper by use of latex. The process is too expensive for the manufacture of toilet paper or note paper but is competitive for napkins, tissue papers and diapers etc. (UNIDO, 1994).

2). Leather Industries
1) Use of ammonia-free CO, deliming for light pelts, combined with the use of ammonium-free bates in the beamhouse;
2) Enzymatic de-hairing of skin and / or de-greasing of leather;
3) Use of biodegradable surfactants instead of organic solvents and their reuse in the tanyard;
4) Use of heavy metal-free and 'benzidine-free' dyes, and non-halogenated fat liquors in the finishing process;
5) Use of water-based finishes. (Sinha and Herat, 2005).

The Germanakos SA Tannery in Greece, used a cleaner technology for recovery of trivalent chromium (Cr^{+++}) from the spent tannery liquors and reused it. Tanning of hides is carried out with basic chromium sulphate (Cr (OH) SO_4), at a pH of 3.5 – 4.0. The tannery liquor is pumped to treatment tank and a calculated quantity of magnesium oxide (MgO) is added while stirring to achieve a pH around 8.0. The chromium precipitates as compact sludge of Cr $(OH)_3$. The sludge is dissolved in concentrated sulfuric acid (H_2SO_4) to get back chromium sulfate (Cr (OH) SO_4) for reuse. In this process about 95 to 98 % of the chromium is recovered. (UNEP and IEPAC, 1993).

3). Textile Industry

1) Bleaching the cotton with 'hydrogen peroxide' instead of chlorine.
2) Use of phosphate-free detergents and CFC-free dry-cleaning agents for washing and cleaning;
3) Use of peroxide instead of chlorine, advanced bi-reactive dyes and enzymes for desizing in the textile dyeing / finishing process.

In the Novotex Textile Company of Denmark only water-soluble dyes are used and chlorine for bleaching is replaced by hydrogen peroxide. In the drying process mechanical finishing is carried out, eliminating the use of chemicals like formaldehyde. The Century Textile Mills of India developed a cheap, non-toxic vegetable product 'hydrol' to substitute the sodium sulfide. 100 parts of sodium sulfide could be substituted by 65 parts of hydrol plus 25 parts of caustic soda. The new technology completely eliminated the emission of toxic sulfides in the effluent. The new clothes were much better in quality. No capital expenditure was involved in the substitution of hazardous chemical as the hydrol is waste by-product of starch industry. It saved them about US $ 12,000 in capital expenses. The savings in not having to install additional effluent treatment facilities was about US $ 20,000. The operating cost also lowered marginally saving about US $ 3,000 every year. (UNEP Reports, 1996-2004).

4). Photography Industry

Photography industry is based on the copious use of chemicals and solvents. Liquid waste from a large-scale film developing and printing operations consists of chromium and spent hypo solutions of developer and fixers, containing 'thiosulfates' and compounds of silver. The silver metal is toxic in certain concentrations. The solutions are usually alkaline and contain various organic reducing agents. The Polaroid Company of U.S. which produces instant photographs drastically cut the use of most of the toxic chemicals and maximized the in-built recycling of the less hazardous chemicals. It completely eliminated the use of category II chromium (Cr) (VI) compounds which were carcinogens and highly toxic, replacing them by recyclable dyes with higher molar absorptivity and hence requiring less dye per picture. (Pollack, 1993).

5). Electronics Industry (Eliminating and Changing Chemicals)

Electronics industry uses several hazardous chemicals including toxic heavy metals (lead, cadmium, mercury, chromium, barium etc.), acids, plastics, chlorinated and brominated compounds in production process. Most hazardous is the cathode ray tubes (CRTs) in the color computer monitor and the TV screens which contain large amounts of lead (Pb) and other toxic chemicals.

1) The electronics industry has also developed over 10 alternative batteries- all free of toxic heavy metals and with better service life;
2) Major computer manufacturing companies are now replacing the hazardous CRT monitor by liquid crystal display (LCD). These flat screen monitors requires much less lead and other hazardous materials;
3) Matsushita is eliminating use of toxic substances from production and has developed first ever lead-free solder, non-halogenated lead wires and non-halogenated plastics;
4) Sony Corporation has developed a lead-free solder alloy, which can be used in conventional soldering equipment;
5) Hewlett-Packard has developed a safe cleaning method for computer chips using carbon dioxide instead of hazardous solvents;

Other changes towards development of clean chemical-free computers are-

1. Printed circuit boards can be redesigned to use a different base material, which is self-extinguishing, thereby eliminating the use of the dangerous brominated flame retardants;
2. PVC plastics used in computer housing and molding is being replaced by ABS plastics;
3. Use of metal shields in computer housings can eliminate use of brominated flame retardants in the plastics;
4. A safer alternative to PVC cabling are low-density polyethylene and thermoplastic olefins;
5. Bio-based plastics, toners, glues, and inks made from plant produced polymers are environmentally friendly products for computers.
6. Researches at University of Delaware in the U.S. have discovered that chemical free computer circuit board can be made from the chicken feather waste. (UNEP Reports, 2000-05)

People's Protest and Campaign Against Hazardous Chemicals in the Consumer Products

In developed nations women are up in arms against chemicals. Naturally because they have been the worst sufferers from it. There is a definite relationship between 'women, chemicals and cancer'. *In her lifetime a woman eats about 2-3 kg of lipstic.*

In 1988, the Women's Environmental Network' (WEN) of U.K. campaigned against the paper industries of UK and their production process, especially focusing on the products that women bought and used such as baby's disposable nappies and sanitary towels bleached by chlorine. At that time, 87 % of the paper that U.K. was producing or importing, was unnaturally white and chlorine bleached. Through the use of such papers U.K. was directly responsible for creating world-wide, up to 90,000 tones per year of a 'chemical cocktail' so lethal that one drop in swimming pool was enough to kill a trout. Within 6 months of launching of the campaign the industries agreed to change their production process and stopped the use of chlorine to bleach their products. Within months environmentally friendly alternatives appeared in the supermarket shelves.

By educating and empowering women about such small consumer products as nappies and sanitary towels, WEN was able to create a climate where general public started questioning the use of toxic chemicals in other paper products, and all such products were boycotted by the consumers thus also reducing the generation of toxic chemical wastes at source. In 1991 the 'Women's Environment and Development Organization' (WEDO) a global coalition of 20,000 women activist based in NY, U.S., called for the recognition of a global, environmentally induced 'cancer epidemic' and demanded for the removal of all 'carcinogenic chemicals' from the products used by the women and children in their daily life.

Genetic Protection from Chemicals in the Environment: Blessing for Mankind?

Scientists at the Bloomberg School of Public Health's Department of Environmental Health Sciences in the U.S. have identified a 'master gene' in mice that controls the action of 50 other genes whose products protect the lungs against environmental pollutants. The master gene named as 'nrf2' is activated in response to environmental pollutants which then turns on numerous antioxidant and pollutant-detoxifying genes to protect the lungs from developing emphysema. The mice were exposed to cigarette smoke. Scientists have identified 50 nrf2-dependent antioxidant and cytoprotective pulmonary genes that work together to protect lungs from cigarette smoke and the automobile exhaust induced emphysema. Dr. Shayam Biswal of the school also indicated that the master gene nrf2 was activated in response to an anti-cancer agent 'sulforaphane'.

Public Opposition Against Nuclear Power Policy to Reduce Radioactive Materials in the Human Environment

Lifecycle analysis from uranium mining to uranium fission in reactors indicates that nuclear power can never be cleaner and safer source of energy for civilization. There has been vigorous opposition from the public and the global intellectual community against nuclear power all over the world. The famous nuclear physicist Sir Brian Flower said in the 6th Report of UK Royal Commission on the Environment – *'The spread of nuclear power will inevitably facilitate the spread of the ability to make nuclear weapons'*. More nuclear would mean more nuclear power – the power 'to kill' and wipe out the civilization before they are struck by the effects of global warming and climate change. It is like substituting one environmental disaster for another.

However, Norway is testing a new type of nuclear fuel called 'yttrium-stabilized zirconium oxide' which will be slashing the amount of 'radioactive plutonium' it creates. It is being tested at the Halden Research Reactor as a replacement for 'uranium oxide', which currently is being mixed with plutonium before irradiation. It would still generate radioactive waste but with a much 'shorter half-life' than the plutonium which has tens of thousands of years.

A 'sustainable renewable energy system' (based on harnessing of mixed sources of hydrogen, solar, wind, tidal and biomass energy) in place of 'fossil fuel and nuclear power' must be established if an environmentally sustainable and socially safe world is to be left to our children and grand-children. Human sustainability and safety on earth demand switching over to the renewable energy systems as soon as possible. Hydrogen fuel is the best bait as it can be used like portable fossil fuel fitting into the existing energy systems of the world but without causing any harm to man and the environment.

CONCLUSION

Relying extensively on 'non-renewable petroleum feedstock' for production of chemicals, conventional industrial chemistry disseminates a 'cocktail of toxic synthetic chemicals' in the global environment, presenting a substantial risk to humans and all living organisms throughout the world. Human health problems resulting from chemical and radiological contamination of the environment is drawing greater attention of the World Health Organization (WHO) than those caused due to microbiological contamination. Exposure to harmful microbes illicit immediate immune reaction from human body that manifest into symptoms like cough, cold, fever etc. indicating the need for quick remedial action. But exposure to chemical and radiation hazards may not manifest into any symptoms for long time and people are not even aware that they have been exposed. Symptoms of exposure may arise several years later, till then the chemicals work as a silent killer in the human body and it might become too late for any remedial action. What is more concerning is that all living organisms on Earth including the human beings have become exposed to chemicals for which there has been no evolutionary experience and hence no biological (immunological) adaptation to cope with as it is in case of microbes.

In the U.S. epidemiological evidence on the 2-3 % of all cancers associated with environmental pollution suggest that exposure to hazardous wastes is a less important risk than exposure to indoor radon and to pesticide residues on foodstuffs. Children specially the infants, are much more susceptible to environmental health problems and at a much greater risk of catching environmental diseases resulting from chemical, radiological and biological pollution of the environment- air, water and the soil. Evidences are gathering which suggest that children in the modern society are becoming more susceptible to cancers. This is more due to their immature immune system that takes time to mature, may be up to 18 years from birth. Children can get exposed to environmental chemicals and radiations at very early stage in their developmental history through mother's placenta. They can also get direct exposure *in utero*. Pre-natal exposure to x-rays, ionizing radiation, pesticides, nitroso compounds in cured meats and solvents have all been linked to higher risk of cancer in children. Paternal occupational exposure to organic solvents like 'benzene' have been found to cause cancer in offspring. Early childhood exposure to pesticides used in the home appears to increase cancer risk in those children. Early exposure to tobacco or biomass smoke, asbestos and ultraviolet radiation increase the chances that those children will develop specific cancers as adults.

Mother's behavior that might help protect the offspring and the children against cancer includes taking multivitamins supplements during pregnancy, breast feeding as long as possible and optimizing the child's dietary intake of fibers, fruits and vegetables. Breastfeeding is critical for transferring immunoglobulins and establishing full functionality of the infant immune system.

After the Stockholm Conference on Human Environment in 1972 and Stockholm Convention on Persistent Pollutants (POPs) in 2001 several nations banned the production of toxic synthetic chemicals through appropriate legislation. Sweden had banned all use of pentachlorophenol (PCP) in 1977 and the Federal Republic of Germany banned all use in 1987. USA cancelled its registration for herbicidal and anti-microbial use and for preservation of wood in contact with food, cattle feed, domestic animals, and livestock. Agricultural use of PCP has been suspended or restricted in among others, Canada, Denmark, German Democratic Republic, and Japan. Canada and the Netherlands have suspended its use for indoor wood treatment. Asbestos use has been effectively banned in Sweden and Denmark. Unfortunately, the US government under those regime refused to sign the Convention on POPs. It assisted chemical industry efforts to undermine the European Union's new unified regulatory system for chemicals. More disappointing is that only 7 % of US federal spending on chemical research and development is devoted to 'green chemicals' and with exception of Greenpeace Society, environmental groups have been slow to take up the cause. However, the American Chemical Society houses the 'Green Chemistry Institute'.

A movement against use of toxic chemicals in consumer products and generation of hazardous wastes, has been going on across the world towards the late 1980s, and women have been on the forefront. For women the world over, breast cancer has become a metaphor for the malignant form of production and consumption of toxic chemicals and destructive development pursued everywhere. The United States and Canadian Government's Joint Commission agreed in 1990 that the onus of proof for chemical safety should be put on manufacturers and users of chemicals, not the general public. In fact, industry intransigence is often aided and abetted by government inaction. A Rather than presuming to keep environment and humans within 'safety limits' and 'acceptable levels' of exposure to toxic chemicals as proposed by WHO, advocates and practitioners of green chemistry aim to produce chemicals that are 'inherently safe' for both man and environment. Green Chemistry Movement (GCM) is going across the world since 1990 to reduce or even eliminate the use of toxic chemicals in industrial production process and search for benign alternatives. GSM in industrial production - reducing or eliminating the use of toxic chemicals and using benign alternatives and chemical processes, is of great significance in all consumer industries. It leads to 'win-win-win' situation for all - benefiting the economy of the industries (not having to spend on costly chemicals and on treatment and disposal of chemical wastes), protecting the environment (not generating any toxic wastes and pollutants), protecting the health of industrial workers and the consumers (preventing them from exposure to toxic chemicals during production and use).

Any progress we make towards sustainable human development hinges upon an honest assessment of the risks and benefits analysis of the chemicals and radioactive materials in the environment, affecting the society and the economy of the nation. The environmental community is calling for a ban on all chlorinated compounds, despite a wealth of research from respected scientific organizations supporting the view that chlorine chemistry can be used safely. Certainly, chlorinated compounds that are persistent, toxic and bio-accumulative (PTBs) needs to be banned. Environmentally sound chemical policy that ensures economic prosperity with environmental security, and also meets all societal goals, is necessary if we are to make progress towards sustainable development.

We are all stakeholders in sustainability- the academia, government, private citizens, the environmental community, business community and the corporate. Without sound science and technology to guide us, our quality of life in the 21st century will worsen. What we need is a way to integrate scientific principles with societal concerns. Human society has much to gain by the careful, yet swift, scientific examination of chemicals to be produced and used in production process. Science has helped us identifying the harmful chemicals specially the carcinogens (causing cancer), teratogens (causing birth defects), immunotoxins (impairing human immune system) and endocrine disruptors (having oestrogenic effects) used in production process and their release into the human environment, making it possible to restrict or even eliminate their production and use.

More works on 'master gene' that controls the action of 50 other genes whose products protect the lungs against environmental pollutants is needed. Manipulation of such 'master genes' through biotechnological researches could become a 'blessing' for mankind to combat pollution related health problems.

REFERENCES AND ADDITIONAL READINGS

Allenby, B.R. and D.J. Richards (eds) (1994): *The Greening of Industrial Ecosystems*; National Academy Press; Washington D.C.

Anastas, P.T. and Warner, G.C. (1998): *Green Chemistry : Theory and Practice;* Oxford University Press.

Anastas, P.T. and Williamson, T.C. (eds.) (1998): *Green Chemistry : Frontiers in Benign Chemical Synthesis and Processes;* Oxford University Press, Oxford.

Anonymous (1994): Potent Immune System Poison- Dioxin; In Rachel Carson's *'Environment and Health Weekly*; Vol. 414, November 3, 1994.

Buzzelli, D.T. (1997): Chemicals : Our Millennium Challenge; *Our Planet*; UNEP Pub., March, 1997.

Christopher, Harris and Harvey Scott (1993): *The Emergency Planning and Community Right-to- Know Act of 1986: Hazardous Chemicals and the Right to Know;* Executive Enterprises Publications Co. Inc., New York; pp. 343.

Curtis, D. Klassen et.al., (1999): *Toxicology*; McGraw Hill Pub., 5th Ed.

EHC – WHO (1991): Partially Halogenated Chlorofluorocarbons (Methane Derivatives); *Environment Health Criteria*, No. 126; p. 197; WHO Publication (1991-1998).

EHC – WHO (1992): Partially Halogenated Chlorofluorocarbons (Ethane Derivatives); *Environment Health Criteria*, No. 139; p. 138; WHO Publication (1991-1998).

Hinwood, A. *et al*, (1999): Cancer Incidence and High Environmental Arsenic Concentrations in Rural Populations (Victoria, Australia) : Results of an

Ecological Study; *International Journal of Environmental Health Research*; Vol. 9; pp. 131-141.

Hoffman, R.E. et. al, (1986): Health Effects of Long-Term Exposure to 2,3,7,8 –Tetrachlorodibenzo-p-Dioxin; *J. of American Medical Association*; Vol. 225: 2031 –2038

Larsen, J.C., P.B., (1998) : *Chemical Carcinogens*; In Hester, R.E. Harrison, R.M. (Eds.) *Air Pollution and Health Issues in Environmental Science and Technology*; Vol. 10; The Royal Society of Chemistry, Cambridge; pp. 33-56.

Matlack, Albert (2001): *Introduction to Green Chemistry*; CRC Press, Washington, D.C.

Nemerow, Nelson. L. (1995): *Zero Pollution of Industry*; John Wiley Pub. NY.

Patton, S. (2004): Toxic Trespass; *Our Planet*; Vol. 15: No. 2; p. 24-26.

Pollack, Susan (1993): *Cleaner Production Makes Money*; Our Planet; Vol. 5; No. 3; UNEP, Nairobi, Kenya.

Rajgopalan, Sanjay and Qinghua Sun (2009): *Ambient Air Pollution Exaggerates Adipose Inflammation and Insulin Resistance in Mouse Model of Diet-Induced Obesity*; Circulation of Ohio State University Medical Center; USA.

Sinha, Rajiv K (1993): Automobile pollution in India and its human impact (A case study on the traffic policemen in Jaipur city of India); *The Environmentalist (Journal of Science and Technology Letters, Middlesex)* UK; Vol. 13 (2): pp. 111- 115.

Sinha, Rajiv K (2007): *Development and The Changing Global Environment*; (p.320); Pointer Publisher, India; ISBN 978-81-7132-498-9

Sinha, Rajiv K (2007): *Sustainable Development* (Striking a Balance Between Economy and Ecology); (p.340); Pointer Publisher, India; ISBN 978-81-7132-499-6;

Sinha, Rajiv K. (2006): *Development, Environment, Human Health and Sustainability*; Pointer Publishers, Jaipur, India; Two Volumes – Vol. I (The Price of Unsustainable Development : p.324) ISBN 81-7132-480-0; Vol. II (The Wisdom of Sustainable Development : p.342) ISBN 81-7132-481-9.

Sinha, Rajiv K. and Margaret Greenway (2004): *Green Technologies for Environmental Management and Sustainable Development*; Pointer Publisher, India; pages 432; ISBN 81-7132-375-8;

Sinha, Rajiv K. and Sunil Herat (2003): *Industrial and Hazardous Wastes: Health Impacts and Management Plans;* Pointer Publishers, India; p.367; ISBN 81-7132-365-0.

Sinha, Rajiv K. and Sunil Herat (2004): Cleaner Production Technologies in Industries With Case Studies- II : Chemical, Paper, Leather, Cement, Textile, Paint and Photography Industry; *Indian Journal of Environmental Protection*; Vol. 24 (8); pp. 561-570; ISSN 0253-7141; Regd. No. R.N. 40280/83; Indian Institute of Technology, BHU, India.

Sinha, Rajiv K. and Sunil Herat (2004): *Cleaner Industrial Production*; *Indian Journal of Environmental Protection;* Vol. 24 (2), pp. 81-89; ISSN 0253-7141; Indian Institute of Technology, BHU, Varanasi, India.

Sinha, Rajiv K. and Sunil Herat (2004): Cleaner Production Technologies in Industries With Case Studies- I : Energy, Fertilizer, Metal and Mining Industries; *Indian Journal of Environmental Protection*; Vol. 24 (4), pp. 241-250; ISSN 0253-7141; Regd. No. R.N. 40280/83; Indian Institute of Technology, BHU, India.

Sinha, Rajiv K. and Sunil Herat (2005): *Cleaner Production : Greening of Industries for Sustainable Development;* Pointer Publishers, India; p. 253 ; ISBN 81-7132-401-0

Sinha, Rajiv K. and Sunil Herat (2005): *Green Chemistry Movement for Cleaner Production in Industries Using Chemicals*; Proceedings of the National Seminar 'CHEMECA 2005', September 25-28, Hilton, Brisbane. (www.icms.com.au/chemeca2005)

Trape, A.Z. (1985) : *The Impact of Agrochemicals on Human Health and the Environment*; In

Trout, P.E. (1972) : PCB and the Paper Industry; *Environmental Health Perspectives*; Vol. 1; pp. 63.

UNEP (1992-2006): *'Our Planet'* (All volumes); Publication of United Nation Environment Program, Nairobi, Kenya.

UNEP (2002): *Cleaner Production Global Status*; Report of United Nation Environment Program (Part A); pp. 1-43.

UNEP and IEPAC (1993): *Cleaner Production Worldwide*; United Nation Environment Program and Industry and Environment Programme Activity Centre, Paris.

UNEP Publication *'Industry and Environment'*, Vol.8; pp.10.

UNIDO (1994) : *Cleaner Industrial Production in Developing Countries;* OECD Workshop on Development Assistance and Technology Cooperation for Cleaner Industrial Production in Developing Countries, Hannover, Germany, September 28-30, 1994; United Nations Industrial Development Organization.

WHO (1981) : *WHO Environmental Health Criteria,* No.18 (On Arsenic); Geneva

WHO (1990) : *Public Health Impact of Pesticides Used in Agriculture*; WHO Publication, Geneva.
WHO (1990) : *WHO Environmental Health Criteria*, No.106 (On Beryllium); Geneva; 210 pages
WHO (1991) : *WHO Environmental Health Criteria*, No.108 (On Nickel); Geneva; 383 pages.
WHO (1991) : *WHO Environmental Health Criteria*, No.118 (On Inorganic Mercury); Geneva; 168 pages.
WHO (1992) : *WHO Environmental Health Criteria*, No.134 (On Cadmium); Geneva; 280 pages.
WHO (1992-2006) : Various Reports of WHO; Geneva
WHO (1993) : *WHO Environmental Health Criteria*, No.140 (On PCBs); Geneva; 682 pages.
WHO (1993) : *WHO Environmental Health Criteria*, No.150 (On Benzens); Geneva; 156 pages.
WHO (1994) : *Some Industrial Chemicals* ;IARC Monographs on the Evaluation of Carcinogenic Risks to Humans; Vol. 60: p.560; Geneva
WHO (1995) : *WHO Environmental Health Criteria*, No.165 (On Inorganic Lead); Geneva; 300 pages.
Woodhouse, Edward J., and Steve Breyman (2005): Green Chemistry as Social Movement; *J. of Science, Technology and Human Values*; Vol. 30.

INDEX

A

abatement, 1
accidents, 28, 95
accounting, 41
acetic acid, 60
acetonitrile, 50
acid, 67, 94
acrylonitrile, 49
acute renal failure, 17
adaptation, xi, 2, 19, 123
additives, 56
adenocarcinoma, 75
adhesives, 12, 50, 51, 60, 61, 114
adipose, 1, 62, 84
adipose tissue, 1, 62, 84
adults, 44, 69, 76, 88, 124
aerosols, 59, 92
African Americans, 30
age, 21, 41, 43, 98
agent, 4, 51, 57, 58, 74, 122
aging, 26
aging process, 26
agriculture, 13, 43
air emissions, 113
air pollutants, 28, 29, 30, 34, 37, 39
air quality, 30, 39
airways, 33
Alaska, 7, 53, 61
albumin, 73

alcohol, 11
alcoholism, 70
allergic reaction, 72
alloys, 70
alternatives, xii, xiii, 103, 111, 121, 125
amalgam, 76
ambient air, 59, 77
amines, 44
amino acids, 115
ammonia, 12, 115, 117
analgesic, 91
anemia, 27, 40, 50, 90
angina, 29
angiosarcoma, 70
aniline, 51
animals, 54, 59, 62, 63, 66, 74, 81, 91, 96, 125
antibody, 86
anti-cancer, 4, 122
antioxidant, xii, 3, 121
antiozonants, 58
appetite, 75
aquifers, 65
Argentina, 94
aromatic hydrocarbons, 23
aromatics, 59
arsenic, 15, 16, 45, 70, 71, 88, 90
arsenic poisoning, 45, 71
arteries, 37, 44
arthritis, 26, 59
asbestos, 15, 22, 39, 71, 87, 107, 124

Asia, 1, 106
assessment, 109, 126
assessment procedures, 109
asthma, 26, 29, 33, 39, 40, 41, 62
atoms, 96, 101
Australia, 31, 61, 94, 99, 114, 115, 116, 129
Austria, 110
automobiles, 12, 74

B

background radiation, 3
bacteria, 43
bacterial infection, 32
Bangladesh, 45
barium, 119
batteries, 12, 27, 72, 74, 76, 78, 116, 119
behavior, 124
Belgium, 94, 110
benign, xii, 103, 110, 112, 116, 117, 125
benzene, 12, 15, 22, 50, 87, 90, 111, 114, 124
benzo(a)pyrene, 40, 63
beryllium, 17, 22, 69, 72
beta particles, 93
beverages, 52, 74
bile, 84
binding, 34, 73, 107
bioaccumulation, 59
biodegradability, 50
biodegradation, 57
biomass, 39, 40, 41, 122, 124
birds, 7
birth, 8, 16, 22, 31, 35, 40, 41, 44, 45, 49, 51, 54, 55, 58, 62, 73, 75, 85, 89, 99, 108, 124, 126
birth weight, 16, 31, 35, 40, 41
births, 75, 98, 99
bisphenol, 46, 63, 88
bladder, 17, 21
bleaching, 11, 55, 117, 118
bleeding, 98
blindness, 40, 60
blood, xi, 2, 32, 34, 35, 37, 43, 51, 56, 61, 63, 65, 73, 74, 75, 78, 80, 83, 90, 99

blood clot, 32, 38
blood flow, 80
blood plasma, 78
blood stream, 32
blood supply, 38
blood vessels, xi, 35, 37, 73, 80, 83
blood-brain barrier, 75
bloodstream, 16, 80, 81
body fat, 21, 53, 84
body weight, 64
boilers, 74
bone marrow, 50, 75, 90, 98
bowel, 27
brain, 2, 9, 16, 21, 27, 29, 35, 37, 46, 63, 74, 75, 76, 77, 78, 88, 89
brain damage, 21, 27, 74, 77
Brazil, 104
breast cancer, 46, 54, 56, 87, 89, 125
breast feeding, 124
breast milk, 56, 62, 63, 81
breathing, 32, 33, 35, 39, 70
brominated flame retardants, 120
bronchioles, 33
bronchitis, 26, 29, 41, 73, 77
butadiene, 111
butyl ether, 3, 60
by-products, 55, 59

C

cables, 62
cadmium, 12, 13, 17, 22, 47, 69, 72, 73, 78, 84, 110, 112, 114, 119
calcium, 44, 73, 98, 117
Canada, 7, 53, 61, 86, 94, 107, 125
cancer, xi, 2, 3, 7, 15, 16, 17, 20, 25, 26, 27, 40, 44, 45, 47, 49, 54, 55, 57, 58, 62, 65, 66, 70, 71, 73, 74, 78, 85, 86, 87, 89, 90, 95, 98, 109, 120, 121, 124, 126
capital expenditure, 118
caprolactam, 63
carbohydrates, 115
carbon, 15, 21, 40, 78, 90, 102, 120
carbon monoxide, 15, 21, 40, 78

Index

carcinogen, 15, 40, 44, 45, 50, 55, 56, 66, 70, 75
carcinogenic potency, 25
cardiac activity, 35
cardiac arrhythmia, 71
cardiovascular disease, 44
case study, 130
casting, 50
cataract, 40
cattle, 96, 98, 124
cell, 16, 17, 32, 62, 69, 77, 78, 86
cellulose, 64
central nervous system, 51, 60, 65, 71, 75
cesium, 92, 93, 94, 98
chemical industry, 110, 125
chemical reactions, 101, 102
Chemical revolution, 1
chemical structures, 19
chemicals causing cancer, 74
childhood, 33, 87, 124
children, 1, 25, 26, 27, 29, 34, 40, 41, 44, 55, 62, 69, 73, 75, 76, 77, 86, 87, 98, 99, 110, 121, 122, 124
chlorinated hydrocarbons, 13
chlorination, 12
chlorine, 11, 12, 46, 55, 76, 117, 118, 121, 126
chlorobenzene, 51
chromium, 15, 17, 69, 74, 110, 117, 119
chronic obstructive pulmonary disease, 40, 73
chronic renal failure, 17
cigarette smoke, 27, 28, 35, 121
cirrhosis, 17
classes, 61, 110
classification, 104
cleaning, 12, 51, 59, 65, 102, 112, 115, 118, 120
climate change, 122
CNS, 60, 64, 66, 76, 77
coal, 25, 34, 63, 70
coatings, 66
Colombia, 41
coma, 35, 52, 60, 64
combined effect, 30

combustion, 32, 78
community, 122, 126
components, 40, 56, 112
compounds, 3, 13, 21, 22, 28, 54, 58, 69, 71, 74, 87, 88, 106, 112, 119, 126
concentrates, 71, 73
concentration, 28, 30, 32, 34, 52, 53, 59, 60, 61, 78, 81, 97, 112, 113
conception, 85
concrete, 96
conditioning, 39
conductivity, 74
conjunctivitis, 40
constipation, 75
construction, 56, 71
consumer goods, 5, 11, 57, 69
consumers, 60, 87, 89, 121, 126
consumption, xi, 5, 11, 55, 76, 98, 125
contaminant, 5
contamination, 9, 13, 23, 47, 54, 70, 98, 99, 123
control, 6, 7, 88, 96, 105, 109
conversion, 93, 115
cooking, 1, 40, 74
cooling, 112, 113
copolymers, 66
copper, 69, 70, 116
corn, 101, 115
correlation, 66
corrosion, 74
cortex, 73
cosmetics, 11, 13, 58, 88
costs, 2
cotton, 114, 118
cough, 123
crops, 47
cyanide, 50, 102
cyanide poisoning, 50

D

danger, 46, 89
death, 26, 31, 33, 34, 52, 60, 61, 64, 78, 81, 97, 98, 100, 108
deaths, xi, 1, 2, 97, 99

decay, 92
decisions, 105
defects, 8, 16, 22, 44, 45, 49, 51, 55, 58, 60, 62, 75, 89, 108, 126
defense, 85, 105
deficiency, 22, 30, 74
deficit, 75
dehydration, 64
Denmark, 110, 114, 117, 118, 125
density, 120
depression, 40, 50, 59, 60
derivatives, 46, 84, 102
dermis, 80
destruction, xiii, 80
detergents, 11, 23, 50, 118
developed nations, 105, 120
developing countries, xi, 1, 2, 20, 29, 39, 47, 105, 106, 108
developing nations, 52
developmental process, 86
diabetes, 1, 29
dialysis, 56
diarrhea, 75
dibenzo-p-dioxins, 51
diesel engines, 32
dietary intake, 124
digestion, 29
dimethylformamide, 16, 17, 57
dioxin, 45, 55, 56, 88
disability, 26
disaster, 56, 95, 97, 122
disorder, 44, 59
disseminate, 102
distillation, 49
distruption of endocrine, 39
division, 70
dizziness, 98
DNA, 16, 50, 74, 77
DNA damage, 74
doctors, 73
dosage, 53, 61
drinking water, 45, 61, 70, 78
drugs, 91
drying, 118
dumping, 9

duration, 21
dusts, 80
dyeing, 60, 112, 118
dyes, 11, 21, 50, 52, 63, 74, 112, 114, 117, 118, 119

E

earth, xii, 9, 45, 94, 95, 111, 122
eating, 80, 81
ecosystem, xi, 2, 19, 81, 92, 99
Ecuador, 116
edema, 78
effluents, 51, 113
elderly, 21, 29, 69, 73
electrodes, 116
electromagnetic, 93
electrons, 26
electroplating, 72
emission, 51, 58, 97, 118
emitters, 94
emphysema, xii, 3, 29, 73, 121
endocrine, xii, 44, 45, 63, 88, 89, 103, 126
endocrine glands, 88
endometriosis, 55, 89
energy, 3, 39, 72, 92, 93, 97, 102, 122
energy efficiency, 102
England, 54
enlargement, 59
environment, xi, xii, xiii, 2, 3, 5, 12, 15, 19, 25, 47, 49, 51, 52, 53, 55, 59, 61, 62, 65, 66, 69, 74, 76, 77, 79, 81, 86, 92, 93, 94, 96, 97, 98, 99, 101, 103, 104, 107, 108, 111, 112, 122, 123, 124, 125, 126
environmental factors, 1
Environmental Protection Agency, 28
enzymes, 71, 83, 115, 118
EPA, 101
epidemic, 54, 121
ethers, 60, 115
ethylene, 16, 57, 111
ethylene oxide, 57
Europe, 97, 109
European Commission, 13
European Union, 5, 103, 106, 109, 125

Index

evaporation, 107
evil, 5, 112
excretion, 78, 83, 84
exercise, 33
explosives, 52, 63, 76
exposure, 1, 2, 17, 20, 21, 22, 23, 29, 30, 31, 33, 34, 40, 41, 45, 50, 54, 55, 57, 58, 60, 61, 62, 64, 65, 66, 69, 70, 71, 73, 74, 75, 76, 77, 78, 79, 81, 86, 87, 111, 123, 125
extraction, 74, 94

F

fabric, 11
failure, 17
family, 54, 116
farms, 12, 54
fat, 53, 117
fatigue, 26, 29, 39, 98
fatty acids, 115
feces, 78, 84
feet, 16, 93, 96
fermentation, 115
fertility, 47, 66, 75
Fetotoxicant, 16
fetus, 16, 35, 60, 63, 70, 75, 81
fever, 73, 98, 123
fibers, 71, 113, 115, 124
films, 58, 66
filters, 66
Finland, 110
fires, 34
fish, 55, 77, 98
fission, 92, 93, 94, 95, 122
flame, 7, 62, 63, 110
flame retardants, 7, 62, 63
flexibility, 12
float, 26
flood, 65
flooring, 109
flour, 59
fluid, 12, 33
focusing, 120

food, xi, 2, 5, 6, 12, 20, 23, 43, 44, 47, 52, 53, 58, 59, 60, 61, 62, 65, 72, 77, 78, 80, 81, 96, 98, 116, 124
food production, 6, 12
formaldehyde, 12, 40, 58, 60, 102, 114, 118
fossil, 32, 78, 122
France, 54, 93
free radicals, 26, 32
fruits, 53, 124
fuel, 3, 92, 93, 94, 122
furniture, 12, 58

G

gamma rays, 93
gasoline, 3, 32, 56, 60
gastrointestinal tract, 59, 91
GDP, 2
gene, xii, 3, 97, 121, 126
generation, 19, 72, 92, 113, 115, 121, 125
genetic protection, 104
genes, xii, 3, 88, 115, 121, 126
Germany, 37, 54, 94, 95, 110, 116, 124, 131
gestation, 75
gift, 114
gland, 88
glomerulonephritis, 77
glycerol, 115
glycol, 12, 60, 114
goals, 126
gold, 17, 69, 70
government, 104, 105, 106, 107, 110, 125, 126
grains, 59, 116
granules, 116
grass, 98
gray matter, 76
grazing, 96
Greece, 117
green revolution, 52, 54
gross national product, 2
groundwater, 45, 62, 81
groups, 77, 102, 108, 125
growth, 44, 62, 64, 69, 76
guidance, 104

H

guidelines, 32
Guinea, 41

hair loss, 62
half-life, 71, 72, 76, 84, 92, 93, 98, 122
hands, 16
harm, 20, 81, 122
hazardous materials, 81, 113, 119
hazardous substances, 109
hazardous wastes, 7, 55, 81, 95, 123, 125
hazards, 21, 30, 66, 109, 123
headache, 29, 34, 35, 64, 73, 98
health, xi, xiii, 1, 3, 19, 20, 21, 22, 25, 28, 29, 30, 32, 34, 43, 47, 50, 52, 54, 55, 59, 63, 66, 70, 72, 75, 85, 88, 89, 92, 97, 103, 104, 107, 108, 109, 111, 123, 124, 126, 127
health effects, 21, 28, 34
health problems, 29, 123, 124, 127
heart attack, 25, 32, 37, 90
heart disease, 29, 31
heat, 30, 51, 115
heat transfer, 51
heating, 34
heavy metals, 6, 15, 16, 17, 47, 69, 81, 110, 112, 115, 119
hemisphere, 95
hemoglobin, 22, 34, 35, 43
hepatitis, 17
hepatomegaly, 71
hepatotoxins, 17
herbicide, 102
heredity, 3, 97
hexachlorobenzene, 86, 108
high blood pressure, 2, 74, 75
high fat, 53
Hispanics, 30
hormone, xi, 63, 88, 89
hospitalization, 34
host, 45, 91
households, 40
housing, 120
human development, 5, 6, 74, 126
human disasters, 24
human exposure, 55, 77
Hungary, 26, 54
hydraulic fluids, 61
hydrocarbons, 32, 33, 40
hydrogen, 96, 117, 118, 122
hydrogen fluoride, 96
hydrogen peroxide, 117, 118
hydroquinone, 58
hypersensitivity, 7, 17
hypertension, 29, 44, 73

I

ibuprofen, 102
immune memory, 85
immune reaction, 17, 69, 123
immune response, 17, 86
immune system, 8, 17, 40, 50, 55, 59, 63, 85, 86, 87, 98, 124, 126
immunoglobulins, 124
Immunotoxins, 17
impairments, 34
impurities, 59
in utero, 124
incidence, 21, 25, 26, 31, 86, 91, 97
inclusion, 88
incomplete combustion, 63
Index, x, 133
India, 29, 40, 41, 45, 53, 54, 92, 94, 98, 99, 112, 115, 118, 130, 131
indigenous, 7, 61, 99
industrial chemicals, 15, 16, 45, 86, 105, 107
industrial processing, 111
industrialized countries, 49
industry, 31, 50, 57, 64, 72, 76, 78, 97, 103, 105, 112, 118, 119, 125
infant mortality, 25, 35, 54
infants, 43, 54, 59, 75, 98, 124
infection, 33, 40, 41
infectious disease, 86
infertility, 13
inflammation, 1, 17, 38
information technology, xi, 6

Index

ingestion, 3, 20, 44, 75, 78, 81, 97
initiation, 71
injuries, 61
insecticide, 116
institutions, 12
insulators, 61
insulin resistance, 1
intelligence, 62, 75
Internation Register for Potentially Toxic Chemicals (IRPTC), 88
international law, 102
International Program on Chemical Safety (IPCS), 103
international trade, 105, 110
interstitial pneumonitis, 77
intestine, 44, 84
intoxication, xi, 2
investment, 114
iodine, 92, 98
ionizing radiation, 124
Iraq, 99
iron, 43, 74, 99
irradiation, 122
irritability, 35
Israel, 54
Italy, 56, 110

J

Japan, 73, 77, 93, 96, 98, 116, 125
jaundice, 17

K

K^+, 64
Kenya, 116, 130, 131
keratin, 80
kidney, 17, 44, 52, 55, 64, 73, 74, 75, 91
kidneys, 17, 29, 45, 59, 70, 73, 77, 78, 83
killing, 41, 60, 112, 113
kinetics, 50
Kuwait, 99

L

labeling, 104
lakes, 62, 99
land, 9, 93, 95, 96
large intestine, 73
larynx, 40
latency, 21
leaching, 45
leakage, 28
learning, 75
learning disabilities, 75
legislation, xii, 103, 105, 110, 124
lesions, 62
leukemia, 56, 70, 90, 97, 98
life span, 69
lifetime, 97, 120
likelihood, 54
linkage, 104
links, 86
lipids, 80
liquids, 80
liver, 17, 27, 29, 55, 56, 59, 60, 62, 64, 65, 66, 70, 73, 74, 78, 83, 84, 91
liver cancer, 17, 27, 66
liver disease, 74
liver transplant, 91
livestock, 125
local authorities, 109
long distance, 7, 61, 107
loss of appetite, 26, 29, 64
low density polyethylene, 115
lubricants, 63
lubricating oil, 52, 60
lung cancer, 3, 25, 27, 28, 29, 31, 40, 71, 72, 99
lung disease, 40, 41
lung function, 33, 41
lymph, 59, 62

M

machinery, 114
magnesium, 118
malaria, 7, 52, 108

malignancy, 97
management, 104, 105, 109
manganese, 69
manipulation, 115
manufacturer, 109, 114
manufacturing, 57, 58, 65, 89, 112, 114, 119
manufacturing companies, 119
market, 109
marrow, 91, 98
maturation, 75
meat, 53, 55, 89
MEK, 60, 61, 112
men, 21, 29, 39, 89
meningitis, 86
mental retardation, 75
mercury, 16, 17, 19, 22, 69, 76, 77, 78, 110, 119
Mercury, 22, 27, 76, 77, 106, 113, 131
mesothelioma, 27, 71
messages, xi, 45, 88, 89
messengers, 88
metabolic pathways, 115
metabolism, 44, 50, 69, 73, 83, 84, 88
metabolizing, 83
metals, 69, 70, 116
metaphor, 125
methanol, 16, 60
methemoglobinemia, 43
methyl tertiary, 3, 60
Mexico, 9, 13, 37, 41, 76
mice, 55, 56, 121
micrometer, 37
microsomes, 77
middle ear infection, 40
military, 72
milk, 2, 53, 55, 62, 81, 95, 98
mining, 2, 3, 70, 72, 92, 94, 97, 122
mitochondria, 77
mixing, 115
modern society, 124
moisture, 80, 115
molecular weight, 23
molecules, 26, 55
monomers, 57
morbidity, 41

morning, 30
mortality, 31, 40, 41
mortality risk, 31
mothers, 35, 40, 54, 59, 77
movement, 125
mucus, 32
muscles, 35
mutagen, 15, 44, 45, 66, 70
mutation, 50

N

Na^+, 64
naphthalene, 63
nasopharynx, 40
nation, 126
natural enemies, 116
nausea, 29, 35, 64
necrosis, 17, 74
Nepal, 41
Nephrotoxins, 16
nerve, 16, 74
Netherlands, 110, 125
network, 102
neurotoxicity, 71
neurotoxicantity, 16
New Zealand, 9
newspapers, 12, 74
NGOs, 108
nickel, 16, 17, 69, 72, 78
nitrates, 43, 44
nitrogen, 33, 34, 40, 43, 44
nitrogen oxides, 34, 40
nitrosamines, 44
nitroso compounds, 87, 124
noble gases, 95
noise, 30
North America, 62
Norway, 122
nuclear weapons, 122
nuclear accidents, 82
nursing, 77
nutrients, 98

Index

O

obesity, 1
OECD, 104, 105, 131
oil, 27, 63, 74, 115
oils, 23
Oklahoma, 96, 99
older people, 75
olefins, 120
ores, 70, 79, 94
organ, 83
organic chemicals, 5, 28
organic compounds, 37, 39
organic matter, 12
organic solvents, 61, 87, 112, 117, 124
organism, 2, 19, 20
Organization for Economic Cooperation and Development, 104
oxidation, 57, 66
oxides, 30, 32, 34, 74
oxygen, 15, 34, 35, 43
oxyhemoglobin, 43
ozone, 33, 65, 112

P

Pacific, 99
Pacific Islanders, 99
packaging, 13, 56, 66, 109
pain, 52, 73
paints, 6, 12, 51, 57, 60, 61, 72, 74, 114
pancreas, 73
pancreatic cancer, 56
paralysis, 16, 66, 75, 77
Parliament, 110
particles, 30, 31, 32, 37, 39, 96
pathogens, 86
PCP, 15, 16, 64, 124
penicillin, 89
peoples protest, 104
perfumes and cosmetics, 12
perinatal, 40
peripheral nervous system, 7
peritoneum, 71
permit, 103
peroxide, 117, 118
pesticide, 20, 44, 51, 52, 84, 116, 123
pests, 7, 52, 53, 116
petroleum distillates, 102
pH, 44, 117
pharmaceuticals, 51, 52, 70
phenol, 64
phenolic resins, 63
phosphorylation, 64
photographs, 119
photovoltaic cells, 70
physical environment, 1
physical health, 29
placenta, 16, 62, 81, 87, 124
plants, 76, 81, 92, 98, 115, 116
plasma, 73
plasticizer, 56
plastics, 2, 6, 12, 50, 55, 61, 62, 63, 72, 88, 115, 116, 119, 120
platinum, 17, 69
pleura, 71
plutonium, 92, 93, 94, 95, 122
poison, 65, 73, 75
Poland, 25
policy makers, xiii
pollen, 32
persistent organic pollutants, 1
pollutants, xii, xiii, 2, 3, 7, 30, 31, 34, 37, 88, 113, 121, 126
pollution, xi, 1, 5, 12, 25, 26, 28, 29, 31, 32, 37, 39, 41, 45, 58, 65, 123, 127, 130
polybrominated biphenyls, 105
polycarbonate, 88
polychlorinated biphenyls (PCBs), 12, 17, 86
polychlorinated dibenzofurans, 55
polymer, 50, 64
polymers, 52, 58, 120
polyurethane, 59
polyurethane foam, 59
polyurethanes, 57
polyvinyl chloride, 55, 56
poor, 1, 25, 30
population, 9, 21, 44, 50, 52, 53, 62, 97, 98, 99

potassium, 98
potato, 53, 115
power, 3, 25, 61, 63, 70, 76, 92, 93, 97, 105, 122
power plants, 3, 25, 63, 70, 92, 97
pregnancy, 31, 41, 124
premature death, 1, 39
pressure, 102
prevention, 91
probability, 32
producers, 94
production, xii, 5, 12, 19, 49, 51, 52, 55, 56, 57, 58, 59, 60, 64, 65, 70, 72, 74, 76, 101, 103, 107, 108, 109, 111, 113, 114, 115, 119, 120, 123, 124, 125, 126
productivity, 47
program, 94
proliferation, 86
propylene, 111
prosperity, 126
prostration, 50
protective coating, 60
proteins, 83
proteinuria, 73
public against nuclear power, 105
public awareness, 3
pulp, 55, 117
PVC, 27, 55, 56, 65, 66, 70, 74, 109, 113, 120

Q

quality of life, 126
questioning, 121

R

race, 21, 93
radiation, 3, 30, 87, 92, 93, 96, 97, 98, 99, 100, 123, 124
Radiation, 97, 99
radioactive isotopes, 97
radioactive waste, 3, 92, 93, 94, 95, 122
radium, 92, 94
radon, 3, 39, 92, 94, 99, 123

rain, 95
rainfall, 96
range, 11, 51, 58, 60, 75, 86, 92, 108, 109
reagents, 101, 102
recognition, 121
recovery, 100, 117
recreation, 98
recycling, 119
red blood cells, 22, 73, 90
refining, 78
reflexes, 62
regulations, 102
rejection, 91
relationship, 54, 120
remediation, 1
renal dysfunction, 60, 73
renewable energy, 122
reprocessing, 94, 96
reproduction, 103
reproductive organs, 3, 99
residues, 44, 123
resins, 52
resistance, 40, 111, 116
respiration, 35, 64, 71
respiratory, 25, 26, 28, 29, 30, 31, 32, 33, 37, 40, 41, 64, 66, 72, 74, 75, 78, 86
respiratory disorders, 33
respiratory failure, 78
respiratory problems, 25, 26
retardation, 44
rhythm, 38, 90
rice, 53, 116
risk, 1, 7, 9, 19, 30, 31, 37, 40, 41, 44, 47, 56, 57, 62, 65, 75, 87, 89, 92, 97, 103, 104, 123
risk assessment, 104
RNA, 77
rods, 94
Royal Society, 129
rubber, 52, 57, 58
rural areas, 39
Rwanda, 116

Index

S

safety, 12, 97, 104, 109, 122, 125
salts, 74, 77, 94
school, 33, 121
search, xii, 103, 111, 125
secretion, 32
security, 126
sediment, 55, 62
seed, 115
sensitivity, 84
sensitization, 17
separation, 93, 101
serum, 78
serum albumin, 78
severity, 20, 21
sewage, 78
shelter, 96
sign, 125
silicon, 116
silver, 119
sinus, 86
sinuses, 26, 29
skeleton, 76
skin, 13, 20, 29, 45, 52, 58, 62, 64, 66, 70, 71, 80, 91, 117
skin cancer, 91
sludge, 78, 118
smog, 26, 33
smoke, 39, 40, 41, 70, 80, 121, 124
smokers, 28, 29, 34
smoking, 3, 35, 37, 41, 70
sodium, 74, 117, 118
soil, xi, 1, 2, 5, 43, 47, 51, 52, 53, 55, 62, 66, 74, 79, 81, 94, 98, 124
solvents, 23, 51, 52, 56, 59, 60, 61, 66, 87, 90, 101, 112, 113, 114, 119, 120, 124
South Africa, 108
soybeans, 101
species, 77
spectrum, 20, 63, 70
sperm, 3, 29, 39, 89
stabilizers, 74
stages, 55, 66
stakeholders, 104, 126
standard of living, 6
starch, 115, 118
steel, 64, 72, 74
stillbirth, 16
Stockholm Conference on Persistent Organic Pollutants (POPs), 92
stomach, 80
storage, 12, 19, 21, 66, 84
stoves, 41
stroke, 32
strontium, 92, 93, 94, 98
styrene, 57
substitutes, 110, 111, 116
substitution, 118
sugar, 115
sugarcane, 116
sulfur, 30, 32, 34
sulfur dioxide, 30, 32
sulfuric acid, 118
summer, 33
Sun, 130
superiority, 72
supply, 12, 35
suppression, 17, 86, 88
survival, xii
survivors, 59
susceptibility, 20, 32, 33
sustainability, 122, 126
sustainable development, 126
sweat, 71, 84
Sweden, 13, 94, 108, 110, 124
swelling, 51
switching, 114, 122
symbols, 104
symptoms, 71, 73, 75, 123
syndrome, 43
synthesis, 78, 101, 115

T

T cell, 86
tanks, 66
Tanzania, 116
tar, 51
targets, 88

technical assistance, 105
teeth, 44
television, 74
temperature, 64, 102
teratogen, 70
terminally ill, 98
territory, 105
testicular cancer, 13, 89
textiles, 2
Third World, 52
threat, 3, 6, 96, 98
threats, 54
threshold, 109
thyroid, 44, 59, 88, 97, 98
thyroid gland, 59
thyroxin, 63
timber, 70
tin, 21, 50, 70
tissue, 71, 83, 87, 97, 117
tobacco, 61, 124
tobacco smoke, 61
Toxic Substances Control Act (1976), 88
toluene, 80
toxic effect, 20, 23, 44, 50, 58, 61, 69, 75
toxic gases, 29
toxic metals, 70
toxic substances, 22, 65, 103, 112, 119
toxicity, 15, 20, 21, 23, 26, 50, 57, 61, 69, 101, 104, 109, 112
toys, 12, 74, 110
trace elements, 31
trachoma, 40
Tratogen, 16
trade, 106
traffic, 37, 130
training, 105
transistor, 76
transmission, 108
transplant recipients, 91
transplantation, 91
transport, 15, 22, 31, 34, 35, 43
transportation, xi, 6, 19
tremor, 77, 96
tuberculosis, 26, 40
tumor, 62, 85
tumors, 58, 62, 70, 75
tumours, 3, 97
Turkey, 59

U

Ukraine, 95, 97
umbilical cord, 37, 76
United Nations, 97, 104, 131
United Nations Industrial Development Organization, 131
United States, 125
upholstery, 65
upper respiratory tract, 32, 51
uranium, 3, 92, 94, 95, 96, 97, 99, 122
urban areas, 30
urinary bladder, 45, 70, 73
urine, 64, 67, 72, 78
USSR, 94, 95, 97

V

values, 34
vapor, xi, 77, 80, 112, 114
vegetable oil, 112
vegetables, 53, 124
vehicles, 37, 57, 74
vessels, 27
victims, 100
Vietnam, 55
vinyl chloride, 16, 55, 56, 66, 111
vision, 34
volatilization, 50, 60
vomiting, 35, 52, 64, 98, 99

W

waste management, 103
wastewater, 12, 112, 113
water supplies, 46
weakness, 64, 73
wealth, 126
weapons, 72
wells, 45

wheat, 53, 59
wildlife, 2, 63
wind, 60, 95, 122
windows, 72, 109
wires, 119
women, 2, 13, 16, 25, 41, 44, 51, 54, 56, 73, 77, 80, 87, 120, 121, 125
wood, 40, 63, 64, 70, 74, 124
workers, 50, 52, 61, 66, 75, 78, 79, 95, 96, 97, 98, 126
workplace, 87

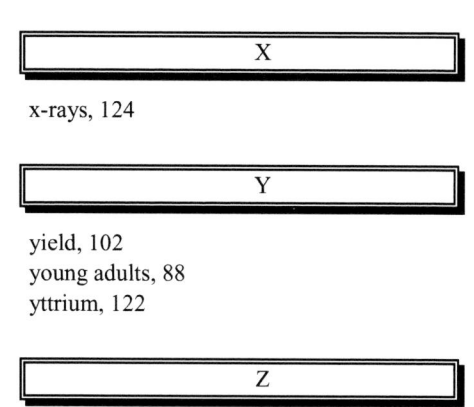

X

x-rays, 124

Y

yield, 102
young adults, 88
yttrium, 122

Z

zinc, 11, 69, 70, 72, 73
zirconium, 122